Astronomers' Universe

Series Editor
Martin Beech, Campion College
The University of Regina
Regina, Canada

The Astronomers' Universe series attracts scientifically curious readers with a passion for astronomy and its related fields. In this series, you will venture beyond the basics to gain a deeper understanding of the cosmos—all from the comfort of your chair.

Our books cover any and all topics related to the scientific study of the Universe and our place in it, exploring discoveries and theories in areas ranging from cosmology and astrophysics to planetary science and astrobiology.

This series bridges the gap between very basic popular science books and higher-level textbooks, providing rigorous, yet digestible forays for the intrepid lay reader. It goes beyond a beginner's level, introducing you to more complex concepts that will expand your knowledge of the cosmos. The books are written in a didactic and descriptive style, including basic mathematics where necessary.

Pierre Léna • Serge Koutchmy

Eclipsed Suns, the Solar Corona and Exoplanets

From Concorde 001 to Telescopes in Space

Second Edition 2025

Pierre Léna
Observatoire de Paris-PSL & Université
Paris-Cité
Paris, France

Serge Koutchmy
Institut d'astrophysique de Paris-CNRS
Paris, France

Translation of Part I by Stephen Lyle

Contribution to Translation of Part II by John Leibacher

ISSN 1614-659X   ISSN 2197-6651   (electronic)
Astronomers' Universe
ISBN 978-3-031-92198-8   ISBN 978-3-031-92199-5   (eBook)
https://doi.org/10.1007/978-3-031-92199-5

Original French edition published by EDP Sciences, Paris, 2023. English title of first English edition was "Racing the Moon's Shadow with Concorde 001"

Photo : Courtesy Ben Cooper, Sequence of pictures of the total eclipse 2024

This Springer imprint is published by the registered company Springer Nature Switzerland AG
The registered company address is: Gewerbestrasse 11, 6330 Cham, Switzerland

If disposing of this product, please recycle the paper.

Serge Koutchmy my friend, the clever astronomer passionate about solar physics, the co-author of this book, died in an accident a few weeks before the book's publication in French, in June 2023 at the age of 82. Having worked at Sacramento Peak in the United States, having collaborated with many partners around the world, having chased so many solar eclipses, Serge was especially keen on making our story available to readers of English. How could I not accede to his wish and not make every effort to accomplish it? Yet, this would not have been possible without the help of Ramon Khanna at Springer, as well as John Leibacher and Jean-Claude Vial, friends and colleagues of Serge, who have been decisive for the existence of this English version of the book. I want to express here my deepest gratitude to the three of them.

I also wish to include in this dedication Mrs. Courtney Ross whose support helped this translation, as well as to the students at her Ross School in East Hampton (United States), for them to happily discover the beauty of science, the richness of human cultures, and how best to serve their sisters and brothers, inhabitants of our planet.

Pierre Léna

Preface

This book is a translation of *Supersonique Concorde 001, couronne solaire et exoplanètes*, published in French in 2023 to celebrate the 50th anniversary of an absolute record, set on June 30, 1973 and never beaten half a century later. On that day, the Sun remained totally eclipsed for 74 min, for the small group of men aboard the supersonic plane Concorde 001 flying in the Earth's stratosphere and thus remaining for nearly an hour and a half in the shadow of the Moon, with many scientific experiments on board.

Part I is a first-person account, since it is told by one of the authors, Pierre Léna, who initiated and lived this adventure at an altitude of 17,000 m, while the second author, Serge Koutchmy, who observed this same eclipse from African soil, had nevertheless installed a telescope on board the Concorde. The former's scientific experiment looked for infrared radiation, emitted by the dust of the solar corona. This theme, in connection with the interplanetary medium and the evolution of the solar system, came to maturity several decades after this historic flight. The other experiments on board the aircraft, including Serge Koutchmy's, focused on another fundamental question of solar physics, although still much discussed in 2024: what is the cause of corona heating and that of the solar wind?

Part II is written jointly by Serge Koutchmy, an astronomer with a passion for the solar corona, and Pierre Léna. It was indeed necessary to show our readers the tremendous progress made during half a century in the knowledge of the corona. This part therefore leaves the style of the story for a more scientific tone, which we have, however, anticipated by some additions and references within the first part. In order to not demand too much from our reader, we have chosen a rich iconography which easily arouses admiration by the beauty of the images, and helps the intuition to enter into the understanding of the complex phenomena which characterize the corona. We have also included some useful concepts in boxes, which the reader may skip during a first reading.

Part II explores what we know about the gas and dust surrounding the Sun and forms its corona, a halo of light for long mysterious because it was only observable during a total eclipse of the Sun, when the Moon's disk masks the intense brilliance of the Sun's. The solar corona remains a fascinating subject of questioning today, studied thanks to a variety of instruments placed on the ground, in airplanes, and especially in space.

With instruments in space, whose technical achievements we did not want to detail, although they are very remarkable and numerous, observation during an eclipse is no longer necessary, because the X-rays and ultraviolet rays emitted by the corona are then observable outside of a total eclipse, and are much more intense than those emitted by the solar disk. In addition, as the Sun's extended atmosphere, it is the link between our star and the Earth, so that the phenomena that take place there, and especially those that are directly directed toward the Earth, have a direct impact on human activities: for example, very energetic jets of particles and magnetic clouds. Finally, eclipses, rare but have become more accessible, are a great opportunity to perceive for oneself the reality of our solar system and to experience an admirable lesson in cosmogony.

The methods and instruments developed for the study of solar eclipses are now finding unexpected applications in the observation of exoplanets around other stars, and especially those of these faint exoplanets, which would be more or less twins of our Earth. This fascinating topic, which focuses on the dusty component of the star's crowns where Earth-like planets are born, is in full development. It therefore deserved to be explored in this book, of which it forms a final chapter, turned toward the future.

Paris, France   Pierre Léna
Paris, France   Serge Koutchmy

Acknowledgments

Our thanks go to those who helped us in the writing and publication of this work, following on the publication of 2014, which was limited to the 1973 flight of *Concorde* 001, and complements it with the immense progress made on the observation and understanding of the solar corona as well as a look turned toward other planetary systems.

Regarding the first part, Pierre Léna expresses his gratitude toward colleagues, researchers, engineers, technicians, administrative staff of the Paris Observatory, the Laboratory of Stellar and Planetary Physics (LPSP) of the CNRS, and the National Institute of Astronomy and Geophysics (INAG) who supported us to design and implement the instruments and then carry out the 1973 eclipse mission, especially the engineers Charles Darpentigny and Alain Soufflot. He is grateful to the many people who helped him relive the details of this adventure in order to write the story: André Turcat first, now deceased, for so many friendly discussions, and Francette Joanne for having authorized him to use André's memories, as well as Michel Rétif, the onboard mechanic; Hubert Guyonnet, radio navigator; and Henri Perrier, test engineer, all too early departed; astronomer colleagues John Beckman, Donald Hall, Donald Liebenberg, Alain Soufflot, Paul Wraight, with whom the enthusiasm of scientific programs during the flight of June 30, 1973 was shared; the whole team of the Air and Space Museum at Le Bourget, including its director, Catherine Maunoury, and its curator, Christian Tilatti, for the installation in 2013 of a permanent exhibition near Concorde 001; Jean-Pierre Sarmant, for an expert and attentive proofreading; and Robert L. Morris, Canadian academic exceptionally passionate about Concorde as much as by eclipses, who provided, with talent, multiple documents and suggestions. The Canadian artist Donald Connolly took extreme care to paint the circumstances of the eclipses of 1912 and 1973, and with great generosity he authorized the reproduction of his two paintings. The images owe a great deal to Jean Mouette, admirable photographer, to Henri Aubry, philatelist and collector, to Vincent Coudé du Foresto who flew onboard the Air France *Concorde* for the 1999 eclipse, to André Girard, from ONERA (National Office for Aeronautical Studies and Research), who led the environmental tests on Concorde, and to Jim Lesurf who was part of the British scientific team. The writing of the story came to life thanks to the generous support of the Treilles Foundation[1] thanks to its welcome for a study stay, during which Emmanuelle Morel d'Arleux and Valérie Dubec were of precious help, and to the welcome later given to the book by my publishers Sophie Bancquart and the Editions Le Pommier.

Finally, the first English language edition, published in 2016 under the title *Racing the Moon's Shadow with Concorde 001* by Springer, benefited from the welcome of Ramon Khanna, as well as the important help of Courtney Sale Ross. The Air and Space Museum of Le Bourget, its director Anne-Catherine Robert-Hauglustaine, and its curator Marie-Laure Griffaton have taken extreme care to maintain the memory of the 1973 eclipse around the *Concorde* 001 which is exhibited, and to prepare the 50th anniversary of the flight for June 30, 2023, as well as honoring this book by featuring the logo of the Museum. The Marseille Astronomy Laboratory and its director Jean-Luc Beuzit contributed to this exhibition. With friendship, Francette Joanne has constantly encouraged us to maintain through this book the memory of André Turcat.

May they all be thanked once again.

Regarding the second part of this book, which is original, Serge Koutchmy first thanks those who helped and/or supported him at the Paris Institute of Astrophysics (IAP), during eclipse observations, since 1968 and in chronological order: A. Lallemand, R. Peyturaux, M. Laffineur, D. Chalonge, E. Schatzman, R. Michard, J. C. Pecker, M. Pick, Z. Mouradian, P. Simon, J. Audouze, A. Omont, B. Fort, J. Heyvaerts, J.-C. Vial, J.-P. Delaboudinière, and J-Y. Daniel. Similarly, for all those who supported him and collaborated internationally, starting in 1970, with the Kiev group led by S. Vsekhsvjatsky, A. Nesmjanovich, V. Ivanchuk, N. Dzubenko, then in Russia, a dear friend G.M. Nikolsky, until 1982, I. Kim then M.M. Molodensky, L. Soloviev, A. Grib, M. Livshits, I. Veselovsky, E. Nikogossian, and finally B. Filippov, a

---

[1] The Treilles Foundation (www.les-treilles.com), created by Anne Gruner-Schlumberger, has a particular vocation to foster and nourish dialogue between sciences and the arts in order to advance contemporary creation and research. It also welcomes researchers and writers in the domain of Treilles (Var, France).

collaboration stopped by the pandemic in 2020 and then by the war in 2022. In the United States, J. Pasachoff, K. Schatten, D. Michaels, G. Newkirk, J. Zirker, R. Smartt, W. Wagner, J. Kuhn, L. November, L. Golub, and W. Livingston have provided unrestricted help in different circumstances. Not without emotion, Serge Koutchmy also wants to thank his collaborators or cooperators, students, and associates of one or more eclipses and experiments on planes or even in space: G. Stellmacher, Ph. Lamy, J. Fagot, P. Coupiac, J.-P. Clavier, R. Caron, J.-P. Zimmermann, Fr. Magnant-Crifo, P. Martinez, M. Sarrazin, A. Adjabshirzadeh, K. Bocchialini, Ch. Viladrich, Th. Legault, M. Loucif, J. Vilinga, Concecao Da Silva, A. Zhukov, C. Delannée, C. Bazin, E. Tavabi, Fr. Baudin, Fr. Sèvre, and N. Lefaudeux, among many others, including in Japan Y. Suematsu and K. Shibata, and in Belgium J. Lemaire and Fr. Clette.

Finally, Serge Koutchmy wishes to pay an emotional and grateful tribute to his dear wife and companion of more than 60 years of shared life, Olga Koutchmy-Belianina.

Special thanks to the European Space Agency, NASA, the European Southern Observatory (ESO), for these beautiful illustrations, free of rights, offered by these public organizations and thus returned to the people, our readers, who funded them through taxes.

Finally, the two authors warmly thank Anne-Marie Lagrange for her review of Chap. 9. The French version owed the support of the Airitage Foundation, with Jacques Rocca, Michael Murphy, and Gaëtan Sciacco, as well as the patient collaboration and advice of France Citrini and Sophie Hosotte, our editors at EDP Science. We thank the translator Stephen Lyle for Part I and the precious help of John Leibacher for checking the translation of Part II.

Pierre Léna

Competing Interests The authors have no competing interests to declare that are relevant to the content of this manuscript.

Contents

About the Authors

Pierre Léna born 1937, is Emeritus Professor at the Observatoire de Paris-PSL and Université Paris-Cité, and a member of the Académie des sciences in France. His scientific work contributed to the conception of the European Very Large Telescope, installed in Chile since 1998 and to the implementation of novel imaging methods on this instrument, namely adaptive optics and optical interferometry, which are allowing many discoveries. As a professor and for many years, he also cared for the training of young scientists for research. His interest for science education led him to co-create in 1995, with the Nobel laureate in physics Georges Charpak and the physicist Yves Quéré, the movement *La main à la pâte*, which developed worldwide to foster school children's curiosity with active learning methods. Interested by the history of science he published in 2024, with the geographer Christian Grataloup, an *Atlas historique du ciel* covering 6000 years of relations between humans and the sky and to be soon published in English.

Serge Koutchmy was born at Le Creusot (Burgundy) in 1941 and died in an accident in Djerba (Tunisia) in 2023, a few weeks before the publication in French of the present book. An astrophysicist, he was Research Director at the Centre National de la Recherche scientifique in France. After studying at the Lomonossov University (Moscou) and getting his PhD from the Université de Paris in 1972, he joined the Institut d'Astrophysique in Paris. A world specialist of the solar corona and its observing methods, his use of the largest ground-based solar telescopes, as well as instruments embarked in space led him to many discoveries. He received the Medal of the Centre National d'Etudes Spatiales (CNES) in 1983. He collaborated with astronomers in United States, Japan, and Russia and organized many expeditions to observe total solar eclipses, including his participation in the 1973 Concorde eclipse flight. As strong supporter of amateur astronomers in the Société Astronomique de France, he received in 1998 its prestigious Janssen Medal.

Part I

74 Minutes of Totality
Onboard Concorde 001

1

Eclipses and Humankind

The total solar eclipse of June 30, 1973, was to be called "the eclipse of the century" (Fig. 1.1). On that very day, the summer Sun rose over Las Palmas, the capital of the Canary Islands off the African coast, but this time there was something different in its appearance. A piece of the solar disk was missing, blacked out by the edge of the Moon, for our satellite had just begun to move between the Sun and the Earth. At sunrise on the same day, far to the west in Dutch Guiana (now Surinam), the tree frogs of the species *Hyla calcarata* began to chorus, despite the unusual time of day.[1] For indeed, it was there on the equator that the great dark disk formed by the Moon's shadow on the surface of the Earth, surrounded by penumbra (Fig. 1.2), began its race toward the east, moving at a speed of more than 2000 km an hour relative to the ground, soon to cross the African coastline. Astronomers, who know how to predict not just the occurrence, but all the details of these eclipses, had explained how this one would be total at every point on the Earth that the shadow would sweep past and partial in each region falling only in the penumbra. Better still, in Africa, total obscurity would last, exceptionally, for 7 min, and the sky would be so dark that the stars would be visible even at mid-day. This record length would make it the eclipse of the century, as announced by the media. By around 10 o'clock in the morning, the black indentation had grown to block out almost half of the Sun's disk. Slowly, a

---

The original version of the chapter has been revised. A correction to this chapter can be found at
https://doi.org/10.1007/978-3-031-92199-5_10

---

[1] Lescure (1979).

P. Léna, S. Koutchmy, *Eclipsed Suns, the Solar Corona and Exoplanets*, Astronomers' Universe,
https://doi.org/10.1007/978-3-031-92199-5_1

Fig. 1.1 The solar corona, photographed by Serge Koutchmy from Moussoro in Chad during the total eclipse of 30 June 1973. The observer is in the umbra, 100 km south of the line of centrality. This implies that the disc of the Moon is not exactly centered on the disc of the Sun. The overexposure of the internal parts of the corona is avoided with the use of an optical filter, becoming darker when getting closer to the Sun edge. (Source: S.K./Institut d'astrophysique de Paris-CNRS)

great white bird began to move across the tarmac at Las Palmas airport to position itself on the runway. Fitted out with a bright red survival suit—standard dress for test flight crew members, in case they need to be fished out of the sea—I found myself aboard Concorde 001 F-WTSS, the prototype of the future supersonic passenger aircraft to be flown by Air France and British Airways, with seven other astronomers. At the controls, André Turcat, the famous pilot who had directed all the test flights of this plane since 1969, and Jean Dabos, both kitted out with the same survival gear. The four other men making up the crew were ready at their posts: "Sierra Sierra, ready for takeoff," announced the control tower (Fig. 1.3). To within a second of the planned

Fig. 1.2 The shadow of the Moon on the Earth, surrounded by the penumbra. The photograph was taken from space by the French astronaut Jean-Pierre Haigneré on board the Mir space station during the total eclipse of the Sun of 10 August 1999, which was visible in France. (Source: J.-P. Haigneré/CNES)

Fig. 1.3 At 10 h 8 min universal time on 30 June 1973, the prototype Concorde 001, registration F-WTSS, took off from Las Palmas airport in the Canary Islands to meet up, somewhere over West Africa, with the Moon's shadow, already slipping rapidly over the Earth's surface. The afterburn (or reheat) of the four engines has been activated to get the 136-ton airliner off the ground. (Source: Jim Lesurf)

time, the beautiful aircraft took off into the trade winds, accompanied by the roar of its four jet engines, flying to meet the shadow of the Moon somewhere above the barren deserts of Mauritania. Less than an hour later, having accelerated to more than twice the speed of sound in the stratosphere, to which it rises in order to attain its cruising altitude, Concorde 001 was about to encounter the Moon's shadow at precisely the place and time identified by our calculations. The accuracy of the rendez-vous was extraordinary, since we reached it to within 1 s and at less than 2 km from the ideal point. The plane's night time navigation lights went on even though it was close to midday local time and the Sun was close to our zenith. Flying in the lunar shadow, which was moving at the same speed as us, the plane would remain in total darkness for 74 very long minutes, while each of the astronomers was getting busy with the instruments, they had brought along to study the Sun and its corona, taking advantage of this unique opportunity to observe for such a long period of time. When we landed at Fort Lamy, today N'Djamena, the capital of Chad in the very heart of the African continent, I was filled with emotion. This dream which, as a young astrophysicist and teacher, I had sketched out barely a year earlier had just been made a reality. No man had ever seen the Sun eclipsed for such a long time, no flight crew had ever carried out such a difficult encounter so faultlessly, no plane had ever provided such a fine observatory for its team of awestruck astronomers. Fifty years on, this record remains unbeaten.

In the Mauritanian sky, as Tuaregs gazed astonished upon this Sun so quick to hide, three parallel stories would finally come together. The first began with humanity which, in all its cultures, had long feared the total eclipse of the Sun, but then sought to understand its cause, and finally to use it scientifically to study the celestial body which brings us light, heat, and life. The second story concerns the extraordinary aircraft that was Concorde, and for which 001 was the first prototype, flying since 1969 and based in Toulouse. So how was this plane deflected from its main objective, that is, the flight tests of a future supersonic passenger plane, to become a scientific laboratory chasing after the shadow of the Moon? Finally, the third story concerns me more directly, since it took me years to become a research scientist, to discover my interest for the Sun and infrared light, to learn the difficult techniques required to fly telescopes aboard aircraft, to build a team that could accompany me in this project, and to dream up this African flight and turn it into a reality.

This is the encounter that I would like to narrate, and it is a true story. The telescopes and the Sun, the supersonic aircraft chasing the shadow, the astronomers, the engineers, and the pilots who set this exceptional record really did exist. This was no video game in which the player travels at will through space and time, nor a comic strip with something fantastical in every picture. It was a collective human adventure which we really experienced. Here I wish to transmit this tale to those who lived through those times, but also to the younger generation in the hope of kindling other dreams and helping them to confront the hard realities of life. I would so much like to encourage them to aim high and excel themselves in new adventures, fulfilling those dreams. But I do have one regret, for my female readers. There are so few women in this story: astronomers, engineers, and crew were all men, and there is nothing I can do about that. However, during my working life, I have had brilliant female colleagues, an honor to astrophysics, and many of my former students have acquired international acclaim, such as Nabila Aghanim, whom I met in Algiers many years ago, now exploring the depths of the past universe, or Anne-Marie Lagrange, world specialist of exoplanets. We shall find her again in the last chapter of this book. Aviation also has its heroines, such as Commander Caroline Aigle, former student at the *Polytechnique*, later fighter pilot, who sadly passed away in 2007, or Catherine Maunoury, twice winner of the World Aerobatic Championships. As director of the *Musée de l'Air et de l'Espace* in Le Bourget (France), which has housed Concorde 001 since 1973, she accepted to install a permanent exhibit on the eclipse, under the wings of the aircraft. And not forgetting space, with Claudie Haigneré, a medical doctor who decided to become an astronaut and spent a total of more than 25 days in space during her various missions. Women are as much needed in science as in exploits of bravery and intellect.

Many glorious and fruitful projects have marked the twentieth century, such as those undertaken by the men and women who went into space, and in particular the Apollo missions to the Moon. They mobilized so many more people, and much greater resources and intelligence than our modest eclipse expedition, and they brought back immeasurably greater scientific results. However, in my own adventure, astronomy and aviation came together in a quite exceptional, and perhaps unique way, that may not be rivalled for some time to come. So, I feel the story deserves to be told.

1.1 500 Million Years in the Sea

The tiny but elegant cephalopod called nautilus[2] has a calcium carbonate spiral shell which develops day by day with the addition of a new growth line. For at least 500 million years, nautili have come up to the surface on the day of the full Moon to spawn, at which point they insert a partition or septum in

---

[2] http://en.wikipedia.org/wiki/Nautilus/.

the series of growth lines. Taking a fossil nautilus, dated using geological techniques, and counting the number of growth lines separating two successive septa, we obtain the duration of the lunar month at the time when that particular nautilus was alive.[3] We thus discover that this duration was shorter than it is today: the Moon is slowly moving away from the Earth, and this increases its period of rotation about its host planet. The distance is increasing at an average rate of 38 m/century, as can be measured very accurately today using laser ranging. The sea tides are caused by the gravitational pull of the Moon (and also the Sun, with a smaller effect) on the Earth's oceans, but the elastic Earth crust is also deformed by the pull of the Moon. The considerable kinetic energy involved in the motion of these huge masses of water is extracted from the gravitational and rotational mechanical energy associated with the masses of the Earth and the Moon. The frictional effects accompanying the motion of the water finally transform part of this energy into a rise in the ocean temperature, at the expense of the Earth—Moon system. The slowing of Earth's rotation around its axis also means a change in its angular momentum, a quantity which physics requires to remain constant in a system. Hence some angular momentum must be taken from the orbital revolution of the Moon around the Earth, slowing this revolution, which means increasing the distance between Earth and Moon (a phenomenon known as tidal acceleration). But when the Moon moves away from the Earth, its apparent size in the sky will decrease. It is thus thanks to a happy coincidence that, for tens of million years, the lunar and solar disks viewed from the Earth's surface have practically the same apparent size when we look at them in the sky.

But scientists like to question "happy coincidences," just in case they hide an unknown phenomenon. In 2014, an astrophysicist from Oxford speculated that, during the Devonian period about 400 million years ago, the amplitude of the tides on Earth, caused by the Moon and the Sun, could present important variations of this amplitude, from month to month. These tide variations would fill and empty the many ponds in which tetrapods were beginning to develop organs to move on solid soil and stimulate their displacements for survival. Natural selection would then favor the ability of these tetrapods to inhabit the soil. Hence, the equal apparent diameters of Moon and Sun on one hand, and the evolution of life on Earth on the other would have been connected[4]!

---

[3] http://www.thegreatbetween.com/the-chambered-nautilus/.

[4] Balbus (2014), https://royalsocietypublishing.org/doi/abs/10.1098/rspa.2014.0263.

When it happens that the Moon, during its orbit around the Earth, passes between the Sun and a point on the Earth's surface, its disk will exactly cover the solar disk and a total eclipse of the Sun will occur. But in a few tens of millions of years, such a precise obstruction of the Sun's light will no longer be possible, so we must take full advantage. Furthermore, if the Moon were to follow a precisely circular orbit around the Earth, then all eclipses would look exactly alike. But its orbit is in fact an ellipse and its distance from the Earth thus varies slightly during the lunar month: sometimes the eclipse will thus be annular, leaving a thin bright ring of the solar surface (the diamond ring) in view, and sometimes it will be total, when the Earth—Moon separation is minimal.

Let us be more precise for a moment. The distance from the Earth to the Sun also changes slightly, since likewise, the Earth's orbit is not quite a circle but another ellipse, even though very close to circular in this case. The apparent diameter of the solar disk as it appears to us in the sky thus varies between 33′1″ and 29′22″, where a single prime indicates 1 min of angle, or the sixtieth part of a degree, and the double prime indicates 1 s of angle. The apparent diameter of the Moon, also on its elliptical orbit around the Earth, varies between 33′1″ and 29′22″. We see therefore that, depending on the positions of the Earth and Moon on their elliptical trajectories at the exact time when the eclipse occurs, the lunar disk will obscure the solar disk to differing extents. And apart from these variable distances, many other factors are relevant to determining the circumstances of an eclipse: its date and time, the geographical locus of observation, and the duration of totality or partiality. This is not the place to go into such detail, but the interested reader may refer to the appendix at the end of the book. It explains why the eclipse of 30 June 1973 was so exceptional.

1.2 The Sun Was Put to Shame and Went Down in the Daytime

In the ancient city of Ugarit on the Syrian side of the Mediterranean, archeological excavation turned up a clay tablet telling of an eclipse of the Sun, perhaps the most ancient ever to be reported.[5] Today, by careful calculation,

---

[5] The earlier eclipse of 21 October 3784 in India may have been reported in a chronicle (Guillermier and Koutchmy 1999).

we can reconstruct the exact date: it was 3 May 1375 BC. We read: "On the day of the new moon, in the month of Hiyar, the Sun was put to shame, and went down in the daytime, with Mars in attendance." Indeed, the brightest stars and planets become visible in the darkness that prevails during totality. Less than a century later, five solar eclipses occurring between 1226 and 1161 BC were reported in China on oracle bones found in excavations at Anyang (Henan Province).[6] Today, the term *ri shi* (日食) is Chinese for "eclipse," but it also means "lost Sun," a reference to ancient myths of dragons devouring the Sun during the eclipse. Drums were beaten, while archers shot arrows toward the Sun to scare the dragon away. And beware the astronomer of the imperial court if he made any mistake over his predictions, for his head would roll. In those days, it could be a dangerous job.

Ulysses may also have observed a total eclipse near the Greek island of Ithaca on 16 April 1178 BC, if we interpret in this way the mention of darkness at noon recounted in Book XX of the Odyssey[7]: "See too crowded with ghosts is the porch, ghosts hurrying down to the darkness of Erebus. Out of heaven, withered and gone is the sun, and a poisonous mist is arising."

But let us leave aside the slow scientific progress in the prediction of the dates and loci of eclipses, a long story that goes back to the Babylonians, the Greeks, the Indians, the Chinese, the Mayas, and the Arabs, passing through the world of the Renaissance and finally arriving at the astounding accuracy of modern computation. And it is thanks to the latter that ancient descriptions of eclipses, dated in the chronicles of the day on the one hand and in a universal chronology by celestial mechanics on the other, that we can situate the historical events of the distant past in a single chronology.

1.3   An Astronomer at the Court of the Son of Heaven

Ancient science and the new science of the Renaissance, the Christian West and the Confucian East, came together in a quite remarkable way at the beginning of the seventeenth century in the context of a solar eclipse. Born in Macerata in central Italy in 1552, Matteo Ricci was a Jesuit priest who set off for India, then China, aboard a Portuguese ship. The only Chinese port then open to trade was the town of Macao, where the Portuguese were extremely active. Ricci learnt Chinese and dreamt of making his way to the heart of the Middle Kingdom, the northern capital of Peking or Beijing. Naturally, his

---

[6] http://history.cultural-china.com/en/183History5571.html.

[7] Baikousis and Magnasco (2008), http://www.ncbi.nlm.nih.gov/pubmed/18557587.

aim was to introduce the Chinese people to the Gospel, but he quickly understood that he must first get a better understanding of this country and learn to love it. After a long period spent studying the Chinese culture, he finally arrived in Beijing in 1601 and soon made friends with the mandarins, the local intellectuals, sharing with them the fruits of European mathematics and astronomy, something he had studied at length before leaving.[8] Before long, became the tutor of the son of the emperor Wan Li, teaching him the science of the Western Renaissance. He produced a Chinese version of Euclid's *Elements*, the great classic of Greek geometry, and going the other way, a Latin version of the *Analects* (*Lun Yu*) by Confucius, then unknown in the West. From geographical knowledge based on accurate measurements, he drew up a map of the world as it was known at the time, in the form of a planisphere (*Il Mappamondo, Complete Map of the Earth's Mountains and Seas*), which he offered to the emperor and which was subsequently disseminated throughout Asia. He even dared to place the Middle Kingdom off-center.

Apart from his geographical knowledge and his mastery of standard instruments for measuring the positions and directions of stars—the astronomical refractor being not yet in use—Ricci had tables called ephemerides, refined by the Portuguese for navigation purposes. These gave the motions of the Earth (diurnal rotation about its own axis and annual rotation about the Sun) and of the Moon (rotation about the Earth during the lunar month). And even before arriving in Beijing, he explained why the partial eclipse observed in Nanchang in the south of China fell short of what was predicted by the emperor's own astronomers. In Ricci's own words[9]: "Many people came to ask me about this: was there an error in the calculation or had other causes reduced the degree of partiality? I took the opportunity to explain how an eclipse of the Sun is not universal, since it can be greater at one place than at another. Hence, it might happen to be greater where the calculation was carried out [in Peking], while being less in this much more central province [Jiangxi]. They were happy with this explanation, because it did not contradict the king's mathematicians, and because the reasons I gave them for eclipses of the Sun or the Moon seemed satisfactory to them. All this was unknown to them, especially the cause of the eclipse of the Moon, which their savants had never identified." Later, in Beijing, the solar eclipse of 11 May 1603 provided him with a fabulous opportunity for scientific demonstration. While the imperial astronomers had predicted the beginning of totality with an error of three quarters of an hour and the end with an error of a quarter of an hour, Ricci's prediction, using his own ephemerids, was significantly more accurate, and this earned him a tremendous

---

[8] Masson (2009). See also (Landry-Deron 2013).
[9] http://www.matteo-ricci.org/Opus/opus23.html.

Fig. 1.4 The tomb of Matteo Ricci, preserved and honored in the former Jesuit cemetery, located in the heart of Beijing in the Party School of the Central Committee of the Communist Party of China (Photo P.L.)

reputation.[10] Another competition was again won for the eclipse of January 16, 1665 by the Jesuit Adam Shall, who had succeeded Ricci. In 2010, exactly four centuries after the death of Matteo Ricci, I visited their tombs with some emotion, in the Jesuit cemetery, preserved today in the Party School of the Central Committee of the Communist Party of China in Beijing (*Beijing shiwei dangxiao*). During the Cultural Revolution (1966–1976), the magnificent stone monument which marks the tomb was buried underground to protect it. This Jesuit astronomer, whose remains were finally laid to rest in a China he had come to love so passionately, had lived three and a half centuries before us, and during this intervening period, eclipse calculations had never ceased to improve, right up to our rendez-vous above Africa on 30 June 1973 (Fig. 1.4).

---

[10] Despite a considerable effort, the present author has been unable to find any record of the residual error in Ricci's prediction for the 1603 eclipse. However, the prediction by the Emperor's astronomers is reported with great accuracy in (Souclet 1732) and Ricci's more accurate result cannot be doubted, however, as attested historically by the reputation he gained thereby.

1.4   A Cuckoo Flies over Paris

It is 1912, at the end of the *Belle Epoque* and before the massacres of the Great War. Six years after the courageous attempts of Clément Ader (1897), the Wright brothers finally overcame in 1903 the difficulties of flying a vehicle that was heavier than air. The French became passionately involved in this new science of aviation. The creators of sometimes strange flying machines exercised their imagination, and often their ingenuity. And then a "diamond ring" solar eclipse was announced. It was to occur over the Paris area on 17 April. No further encouragement was needed to inspire two young men, Michel Mahieu, already an accomplished pilot at the age of 20, and Gaston de Manthé, his navigator, to take to the airs and view it from their military-style Voisin biplane, specially built for Mahieu. The plane took off from the landing strip at Issy-les-Moulineaux, crossed the Seine close to the Eiffel tower and, at an altitude of about 250 m somewhere near Saint-Germain-en-Laye, its two passengers encountered totality on the line of centrality during the short transit of the lunar shadow. Their flight lasted 50 min, so they were able to observe totality at 12 h 20 and land 10 min later at Issy-les-Moulineaux. The next day, the newspaper *Le Matin* gave homage to these two men "enchanted by having been able to observe the eclipse from a little closer than ordinary mortals," and above a light mist. The 18 April issue of *Le Matin* published close to a million copies containing an exceptional photo taken by an anonymous photographer. In it, we see the silhouette of Mahieu's Voisin biplane next to the eclipsed Sun. Despite the poor quality of the reproduction, we may surmise that it was the diamond necklace that caught the photographer's eye through the mist. This ring of bright beads around the Sun is observed when the apparent size of the lunar disk is exactly such that the Sun's light can just reach us through the valleys between the lunar mountains standing out from the rim of the Moon's disk (Figs. 1.5 and 1.6).

When the Great War broke out, Michel Mahieu gave his plane to the nation and then left to fight. As a captain, awarded the *Croix de guerre*, he was mortally wounded in the Somme at the age of only 26 years. An impressive statue of him and his brother August was put up in his home town of Armentières. Although he had no scientific pretensions, Michel Mahieu nevertheless made the first airborne observation of an eclipse, and he did it in France. Mahieu's Voisin had a wingspan of 15.75 m, and Concorde, in which we were to fly 60 years later, less than twice that (25.60 m). So Mahieu was our precursor, and it seems right therefore to tell his story here, as brought

Fig. 1.5 The first flight in history (1912) to observe a total eclipse of the Sun. The Voisin biplane was photographed from the ground in the direction of the eclipsed Sun. Because of an unclear sky, the bright corona ring is fuzzy on this poor quality reproduction of the daily newspaper *Le Matin*, April 18, 1912, knindly pointed out by L. Robert Morris

Fig. 1.6 View by Canadian artist Don Connolly of the Voisin plane flown by Michel Mahieu as it crossed the Seine river to join the line of centrality of the eclipse over Saint-Germain-en-Laye. Insert: Parisians watching the eclipse (Source: Don Connoly (Painting *The Birth of Aviation Astronomy: Solar Eclipse* (Paris, April 17 1912). Design by L. Robert Morris. Painted by Don Connolly & L. Robert Morris. Acrylic on board (61 × 46 cm), 2012))

together by Robert Morris, a Canadian teacher and engineer himself fascinated by eclipses and aircraft![11]

Seven years later, during the total eclipse of 29 May 1919 observed in Principe off the coast of Gabon, 2000 km south of the path that Concorde would follow half a century later, the observations were used to check the calculations of Albert Einstein, based on his general theory of relativity published in 1915. Throughout the whole of history, this was without doubt the most fruitful eclipse for the progress of science.[12]

1.5   The Mystery of the Solar Halo

From the nineteenth century, eclipses began to interest astronomers for another reason. Up until then, the aim had been to make ever more accurate predictions of the time and place, thereby putting to the test calculations of the lunar and terrestrial motions. But now it was the inner workings of the Sun itself that eclipses could throw some light upon. The Greeks had already noted the presence of a weak halo of light which became visible around the eclipsed star. Describing the total eclipse he observed in 71 or 83 CE, the historian Plutarch wrote: "In eclipses [of the Sun], the Moon allows a part of the Sun to spill over, and this reduces the darkness." This "part," known today as the solar corona, was of great interest to astronomers, but it would long remain mysterious. Observing this halo in 1567, Johannes Kepler attributes it to the refraction of solar rays by a lunar atmosphere. The great British astronomer Edmund Halley, after the 1715 total eclipse in London, proposes the same explanation. During the short moment of totality, he attempted to measure the displacement of this halo, to see if it would have the same displacement as the Moon. But the ruler placed in the eyepiece of his telescope gave him a biased measurement which led to his misinterpretation. Following Halley, the French François Arago even proposed the name "lunar corona." Arago calls for future observers of the 1842 eclipse, visible in Southern France, to make careful observations. During this eclipse and the one in 1851, several observers, located at distant sites along the totality band, confirmed that the coronal structures belonged to the Sun and not to the Moon. Drawings, and

---

[11] Morris (2013).

[12] Dyson and Eddington (1920). http://www.jstor.org/stable/91137. See https://en.wikipedia.org/wiki/Eddington_experiment.

later photographs of the white corona,[13] confirm that at least an important part of this white corona belongs to the Sun. Nevertheless, it remains that this part is buried in another source of light of yet unknown origin, as noticed and reported by the French astronomer Jules Janssen, observing in India the total eclipse of 1671. This additional component of the corona shall later be identified as the F corona.

During the 1860 eclipse in Spain, the Italian Jesuit astronomer Angelo Secchi definitely established that the fine protuberances—or movement of matter—then observed at the edge of the eclipsed disc are indeed solar phenomena. It is only later, thanks to the spectra obtained by Jules Janssen observing in Siam in 1868 then in India in 1871, that the faint luminous halo observed around the Sun during total eclipses shall be properly attributed to the Sun itself and will take its current name of "solar corona."

> **Box 1 A Short Presentation of the Solar Corona and Zodiacal Light[14]**
>
> The solar corona is an irregular, time-varying, and very faint extension of the Sun's bright disc, which is seen from Earth at an angle of half a degree, whereas the corona extends over several degrees or more. The total brightness of the corona is about a million times less than that of the disc, and less than that of a full Moon seen from Earth at night. Compared with the brightness of our atmosphere, whose molecules and dust scatter sunlight, in daylight not far from the solar edge, the brightness of this corona is a thousand times weaker. The corona can therefore only be observed by terrestrial observers when the Sun, and therefore the Earth's atmosphere that it illuminates, is "turned off" during a total eclipse, or to a limited extent when an "artificial eclipse" is produced by an appropriate optical device, called a coronagraph, which we will present later. Today observations from space provide new insights on the corona seen at ultraviolet and X-ray wavelengths of light.
>
> The corona is made up of a plasma of hydrogen and helium—atoms that are split into electrically charged nuclei and electrons—at a temperature of around one million degrees, and very fine solid particles (silicates) heated to a few hundred

---

[13] Photography introduced more objectivity in science, as testimonies or drawings could always be questioned as reflecting an observer's bias. Personal communication, courtesy of Avice et al. (2024). "Objectivité dans la représentation des éclipses au milieu du XIXᵉ siècle."

[14] This short-framed text allows our readers to catch the main properties of the solar corona, as it was known in 1973. In Part II of this book, Chaps. 7 and 8 will present a more complete description of our knowledge in 2024, as well as the remaining questions.

degrees by radiation from the Sun's surface. The coexistence of these two components of the corona, called K and F, respectively, at such different temperatures is due to the very low density of the corona, which does not allow the temperatures to be equalized. The light emitted by the corona, visible to the naked eye during a total eclipse, is the superposition of the emission from these two sources: the electrons in the plasma scatter the solar radiation, forming the K component of the corona; the solid particles (dust) also scatter it, forming the F component. The brightness of the inner corona, close to the solar surface, is dominated by the first K component, while beyond it the second F component dominates (Fig. 7.3, Part II). Other wavelengths of light (X-rays, ultraviolet, infrared, and radio waves), distinct from those perceptible by the eye, are emitted by the corona and make it possible to study it in greater depth, in some cases outside a total eclipse.

Observing with an instrument, it is possible to distinguish between these two components by the properties of the light they emit. K light is the same white color as the light emitted by the Sun's surface, it is polarized and its spectrum, i.e., the light distribution according to wavelength, is almost continuous. In contrast, light from the F corona, which is not polarized, contains the same spectral lines as those known as the Fraunhofer lines. These lines are also seen in the light emitted by the solar surface, hence the name F.

The corona is sometimes referred to as the Sun's "outer atmosphere." This usage should not lead to believe that what is commonly called the surface of the Sun, which marks the limit of the disc observed in the sky, is a transition between a solid body and a gaseous atmosphere. The Sun is entirely gaseous, and the very distinct appearance of a disc that we see is due to the constant decrease in the density of this gas from the interior of the Sun to the corona. When the density becomes sufficiently low, the light can escape into space, giving the appearance of a well-defined sphere with an edge. The corona is the gaseous part above this surface until far away.

The zodiacal light is an extension of the corona, which can be observed at night under very transparent skies, and more easily at low latitudes, before sunrise or after sunset, away from the parasitic lights of cities. It is a lenticular glow, centered on the Sun and aligned along the ecliptic. The ecliptic is the line of the celestial sphere along which the Sun, as seen from Earth, apparently moves throughout the year, passing from one constellation to another—the constellations of the Zodiac, from which the name zodiacal light derives. Descartes and Newton wondered about the origin of this light, but the breakthrough came with the astronomer Jean-Dominique Cassini, the first director of the Paris Observatory appointed by Louis XIV. In 1683, Cassini noted that the appearance of this elongated area was identical whether seen from Paris or the South of France, and deduced that it was located far from the Earth. With remarkable intuition, he also suggested that it was grains of matter scattering the Sun's light that caused the zodiacal glow. A little later, the astronomer Jean-Jacques Dortous de Mairan (1678–1771) noted a relation between the Sun's activity, observed by the greater or lesser number of sunspots over time, and the brightness of the zodiacal light. He therefore considered that it was "merely the atmosphere of the Sun, a rare and very tenuous fluid or matter, luminous by itself, or chiefly illuminated by the Sun's rays." This explanation was later confirmed and zodiacal light is now understood to be the scattering of the Sun's light by small grains of dust (Figs. 1.7 and 1.8). We shall now focus on the corona itself, while Chaps. 8 and 9 return to this zodiacal cloud, which is of major importance and still not fully understood.

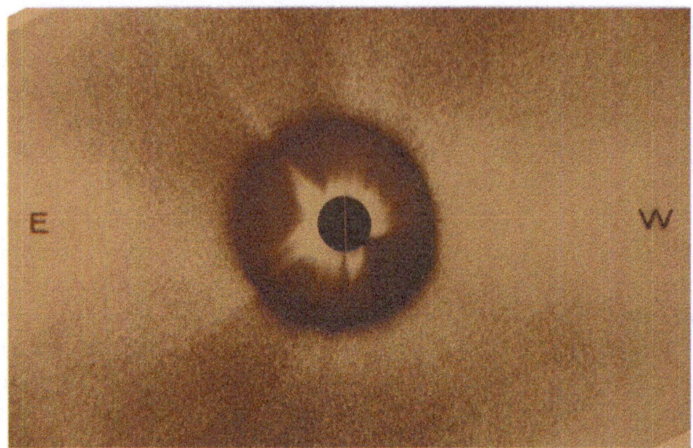

Fig. 1.7 Distant F corona, with reinforcements detected on the solar equator (East-West, E-W) during the eclipse of 7 March 1970 in the USA (maximum solar activity). This unusual image is a photograph taken from the ground in the very near infrared. The black disc is produced by an outer mask that hides the bright inner parts of the corona, dazzling the outer part. Superimposed on this disc is a second photograph of the inner corona K, with its structures. (Source: C. Lillequist & E. Schmahl, High Altitude Observatory, Boulder, Colorado)

1.6   Tourism for Amateurs and Science Above the Clouds

Following the success of Michel Mahieu's flight in 1912 and the ensuing rapid development of aviation, the advantages of observing total eclipses from craft that could fly ever higher soon became a major motivation either simply to experience the intense emotions procured by such a spectacle or to penetrate the mystery of this fleeting solar corona.

On 10 September 1923, in California, the US Navy flew 16 planes, including one seaplane, to determine as accurately as possible the central axis of the path followed by the lunar shadow during a total eclipse of the Sun. One of the pilots was a certain Albert W. Stevens. In 1937, at which time he already held the world altitude record in a balloon, he took an airplane Douglas DC-2 up to almost 9000 m above the Peruvian Andes, photographing for the first time this solar corona whose origins were still so enigmatic.

In 1941, the war was raging in China with the Japanese, and some astronomers had taken refuge in the mountains of Yunnan and its capital Kunming, far in the South-West. A total solar eclipse was foreseen on September 21, 1941 and despite the situation, they managed to send an expedition at the optimum place in the totality path, near Lanzhou (province of Gansu, NW of China). In addition, a Tupolev aircraft bomber, made available to Chiang Kai-shek air force by the Soviets, could fly 9000 m high and photographed the eclipse (Fig. 1.9), taking a movie which became known in USA and contributed to a positive image of the fighting Chinese.[15]

---

[15] Documents kindly provided by Yi Zhou and Thierry Montmerle, former General Secretary of the International Astronomical Union (IAU). See https://eclipse-history.wp.st-andrews.ac.uk/extraordinary-patriotism-demonstrated-by-chinese-eclipse-science-apostles-during-the-1941-total-solar-eclipse/.

Fig. 1.8 The zodiacal light, with the planet Venus on the right above the horizon, is seen from ESO's European observatory in La Silla (Chile), whose telescopes are in the center and on the left. The lenticular shape of the zodiacal light, elongated almost vertically along the ecliptic, has sometimes led to it being called the "pyramid of the sky." Before sunrise, the horizon is surmounted by a pink color. Long called "false zodiacal light," this color is due to the fluorescence of oxygen atoms in the Earth's upper ionosphere, around 300 km altitude, already illuminated by the rising Sun (Source: ESO/B. Tafreshi, twanight.org)

Fig. 1.9 At return of the eclipse observing mission, a young astronomer emerges from his seat in the cockpit of this two-engine bomber Tupolev TU SB. The text in Chinese indicates: *I took the pictures from the aircraft*. The Sun, still partially eclipsed, is reflected by the cockpit shield. (Source: Purple Mountain Observatory, Chinese Academy of Sciences)

Airborne observations were taken further after World War II, in 1945 and during the following years. On 30 June 1954, two saros[16] periods before the eclipse to be observed from Concorde, the British astronomer D.E. Blackwell set to work at an altitude of 9000 m aboard a propeller-driven Lincoln bomber. Indeed, he observed the sky with the door of the aircraft open and in the greatest discomfort, close to passing out, but without the porthole getting in the way of the light from the corona. His remarkable observations mark a step forward. The following year, the French astronomer Raymond Michard successfully made observations of the solar corona during the eclipse of 20 June 1955 while flying aboard a refitted Nord 2501 military transport aircraft in Indochina. For the eclipse of 20 July 1963 in the Canadian north, NASA flew a Douglas DC-8 powered by four jet turbine engines at close to 1000 km an hour, while up to then slower, propeller-powered aircraft had only been able to gain a few seconds by pursuing the Moon's shadow. In 1965, the same NASA team fitted out a Convair-990 plane, Galileo I, for astronomy and eclipse observation, while the US Air Force did several eclipse expeditions from 1965 on (Fig. 1.10). We shall return to these planes later.

---

[16] A saros period is about 18 years, a characteristic interval between solar eclipses repeating at similar locations. See the Appendix 1 at the end of the book.

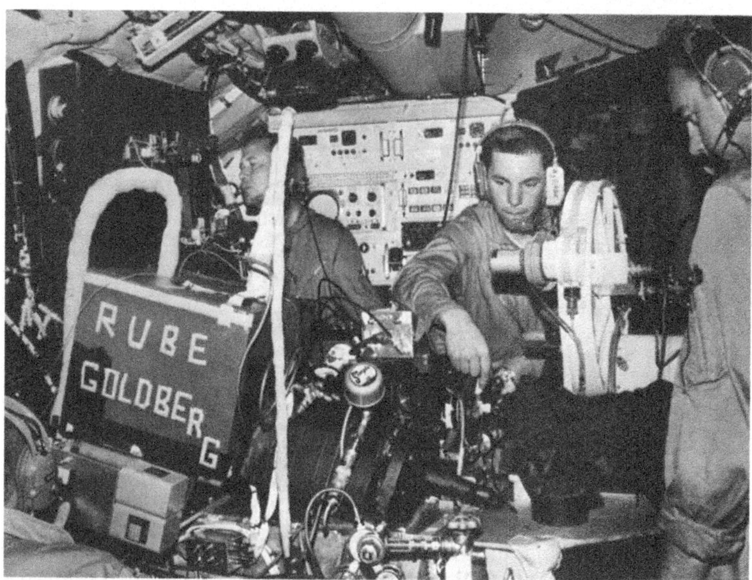

Fig. 1.10 Above Samoa islands in the Pacific in 1965, observation of a total eclipse of the Sun by the team from the Los Alamos Laboratory in New Mexico. The scientific equipment is set up in the cabin of a four jet Boeing NC-135 of the US Air Force, the Sun being viewed toward the left through a specially made porthole. In 1973, Donald Liebenberg (center) was one of the foreign scientists invited to take part in the Concorde flight (Source: Don Liebenberg/Los Alamos Laboratory)

The advantages of these airborne solar observations are clear enough. Any point on Earth can be reached by plane, wherever totality happens to occur. At higher altitudes, the probability of cloud cover is significantly reduced, the Earth's atmosphere is much more transparent to celestial light, and the sky is darker, whence the corona stands out better. Finally, the duration of the eclipse can be slightly increased thanks to the speed of the aircraft as it follows the shadow. But there are disadvantages too, since the light must in principle pass through a porthole which seals the inside of the plane off from outside, while vibrations and the motion of the aircraft can disturb the stability of the onboard telescope used to provide an image of the Sun (Fig. 1.11).

These first flights, often spectacular, led to an improved although still rudimentary understanding of the solar corona and its inherent physical mechanisms. They would thus supplement the many expeditions to observe eclipses from the totality region on the Earth's surface.

Among these, let mention the Soviet vessel Alexander Griboedov which, offshore the Brazilian coast, was carrying a radio telescope aiming at 1.5-m wavelength in order to observe the total eclipse of May 20, 1947. The cloudy

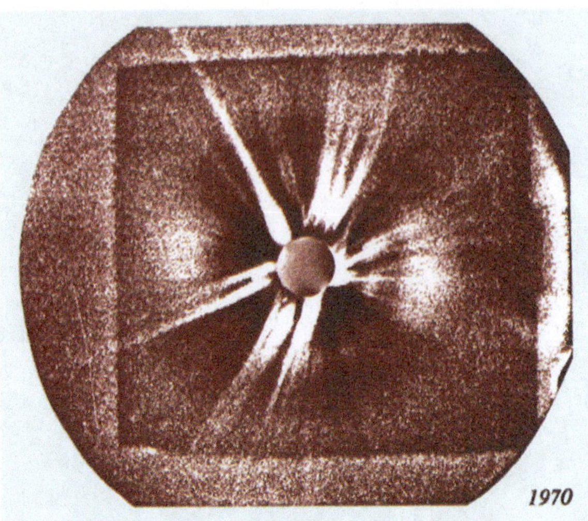

Fig. 1.11 A white light image of the corona, taken during the eclipse of March 7, 1970, from a NC-135 aircraft (Los Alamos Laboratory) above USA. Two reinforcements of light are clearly visible in the equatorial plane (nearly horizontal), belonging to the F corona. The great jets of the K corona are extending in all directions. A numerical treatment avoids the overexposure of the internal corona. Despite its mediocre quality, this historical document deserves to be reproduced Dolci (1997) (Source: Don Liebenberg/Los Alamos Scientific Laboratory)

sky did not allow to see the visible light, but did not hamper the radio waves and a coronal signal was measured. Since the 1940s, the Russian physicist Iosiv Shklovsky (1916–1985) had proposed that the coronal gas was at a very high temperature, without being able to demonstrate it. He was present during this Brazilian eclipse and properly interpreted the radio signal, remaining present even during totality, as a solid proof of an extended emission, around the Sun, of a coronal gas reaching a million of degrees.[17]

---

[17] Rudniskij (2006), https://www.tandfonline.com/doi/abs/10.1080/10556790601165700.

2

Concorde: A Dream Takes Off

While astronomers sought enthusiastically to unravel the secrets of the Sun, the stars, and the universe in general, other men and women dreamt of building machines that could tear themselves away from the Earth's surface and take to the skies, going ever higher, ever faster, and ever further. These were the heroes of the twentieth century, inspiring us even before the first men went into space (Yuri Gagarin in 1961), then to the Moon (Neil Armstrong and his colleagues in 1969), leaving the cradle of humanity to explore new, still more chimerical horizons.[1] The supersonic commercial plane Concorde was part of this saga of the last century, in which the ambition of intercontinental travel for a hundred passengers, putting New York at only three and half hours from Paris, would finally be realized. And it was precisely in this year of 1969 that the Concorde prototype 001 made its first test flight from Toulouse airport in Blagnac, 2 months after its Soviet rival, the Tupolev Tu-144, took to the air for 38 min near Moscow. At this time, the United States project of a commercial supersonic airliner was practically abandoned.

2.1 Entente Cordiale and a Challenge

The decision to build a supersonic passenger plane was not an easy one to reach. There had been two jet passenger planes since World War II, exploiting the technological progress stimulated by this battle of giants. The first was the

---

[1] Bonnet (1992).

British four-jet Comet, which took off in 1949 and was in service from 1952. Two tragic accidents marred its commercial career, leaving the way open for Boeing's long-haul four-jet 707. In Canada, a four-jet passenger plane, the Avro C102, a contemporary of the Comet, was also developed, but it was abandoned in 1951 in favor of a supersonic fighter plane. Other four-jet passenger aircraft would come later: the Vickers VC-10 in the United Kingdom and the Tupolev TU-104, the Ilyoushin IL-62, and IL-86 in the Soviet Union. At the same time, the French Caravelle, a twin-jet aircraft with a quite different design, was extremely successful. Its first test flight was in 1955 and it was used by Air France from 1959.

At the end of the 1950s, military aircraft were just beginning to explore the supersonic range, exceeding the speed of sound, some 1000 km an hour, in the atmosphere, although needless to say, breaking the sound barrier raises serious problems for the aerodynamics of an aircraft. At the height of the Cold War, the US military had a B-58 bomber that could fly at Mach 2, that is, twice the speed of sound, while carrying an atomic bomb. When this supersonic four-jet aircraft traveled non-stop from Fort Worth in Texas to Le Bourget in France in 1961 in just 3 h and 19 min, Lindbergh's maiden transatlantic flight was much in mind. But although it flew at Mach 2 for some of the time, the aircraft had to be refueled in flight twice over the Atlantic. A week later, this plane crashed at Le Bourget during a flight demonstration and its three crew members were killed. There were many losses among the hundred and 28 B-58 aircraft and their crews, as 2 years later only 95 of those planes were still flying.[2] Already engaged in the race to the Moon, the USA also wanted to take its place in supersonic passenger transport, but it failed, and the project of a Boeing 2707 (SST) was abandoned in 1971.

In France as in the United Kingdom, the design offices were also pursuing the possibility of a high-speed commercial aircraft. In France, a four-jet project called Super Caravelle was begun in 1961 and a model was presented at the Paris Air show in Le Bourget. The plane would have carried a hundred passengers over distances up to 4500 km, still not enough to get from Paris to New York, at a speed of Mach 2.2, that is, more than twice the speed of sound. As one can easily imagine, transporting bombs at Mach 2 with a potentially disposable military crew bears no relation to the reliability required for a commercial airliner carrying men, women, and children.

---

[2] Kenneth (1982).

General de Gaulle, whose ambitions for France were well known, realized that France alone could not cover the whole cost of developing such a plane. In November 1962, a few weeks after refusing to allow the United Kingdom to enter the European Common Market, France signed an agreement with its British partners to build a supersonic aircraft for commercial use.[3] This would become Concorde, and not Concord, after long and arduous negotiations about whether to use the English spelling with no "e" at the end. In France, the company Aérospatiale was made responsible for the program, while André Turcat, pilot and engineer from the *Ecole polytechnique*, was chosen to direct flight tests. In 1959, Turcat had beaten the world speed record over a closed loop of 100 km in a ramjet-powered Griffon II aircraft.[4] The ramjet is the simplest conceivable jet engine. The gases produced by the burning fuel are simply ejected at the outlet of a tube and this produces the thrust. No moving parts are required, but this jet engine cannot be used for take-off.

2.2 Concorde Takes Off

There were two prototypes of Concorde, one in France, the other in the United Kingdom, thus sharing the work of exploring the many technical issues that needed to be solved: materials, wings, engines, landing gear, safety systems, flight procedures, and so on. And this was no small challenge (Fig. 2.1). The military aircraft then in service and able to break the sound barrier fell far short of the range, the payload, and the long duration of supersonic flight assigned to the project. The new plane would have to fly in normal engine regime without the afterburn[5] used by military aircraft for a few minutes to increase the thrust of the jet.

André Turcat and others have told the story of the preliminary work. On 2 March 1969, under the eyes of General de Gaulle who was watching on television, the prototype, a beautiful white bird, took off from Toulouse at a speed of 280 km/h (150 knots) with André Turcat and his copilot Jacques Guignard at the controls, while the flight test engineer Henri Perrier and the flight test mechanic Michel Rétif watched over the four engines and the 136 tons of

---

[3] The birth of the Concorde project and the negotiations between the two countries are described in Kenneth (2002). The book contains first-hand accounts by André Turcat and Henri Ziegler.

[4] http://en.wikipedia.org/wiki/Ramjet.

[5] Afterburn or reheat is a fuel-costly method for increasing the thrust of the jet engines for a very brief time by injecting fuel directly into the exhaust gases, where they are set alight. See the Appendix 1 on Concorde at the end of the book.

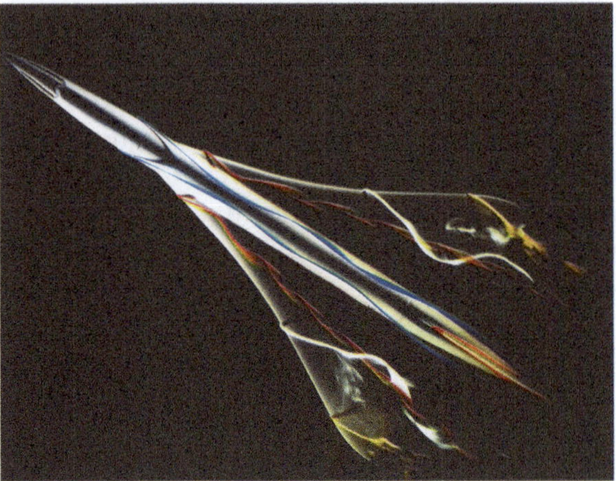

Fig. 2.1 The design of Concorde wings resulted from deep aerodynamical studies, either by numerical simulations or physical models placed in a wind tunnel at the *Office national d'études et de recherches aérospatiales* (ONERA). Here the model, in landing configuration, produces air flow changes, which are visualized by colored liquids emitted at different points. Their trajectories reveal the interactions between the plane and its fluid environment, e.g., the production of turbulence (Source: ONERA)

machinery. Turcat writes: "For us, it was as though we barely spent a moment in introspection, for all our senses were straining over the instruments … There was no room for apprehension, only action; to the point where I wouldn't even experience the usual primitive pleasure of take-off, with all the huge power held in one hand operating four throttles." We will meet up with three of these men 4 years later during the eclipse. The day after this first successful flight by Concorde, NASA astronauts sent up a rocket for the Apollo IX flight, during which they would test all the equipment of the lunar landing module in orbit around the Earth prior to the successful lunar mission of Apollo XI. Thus, in his brief and modest speech after landing before an enthusiastic crowd in Toulouse, André Turcat sent them warm congratulations. On either side of the Atlantic, two of the most unlikely adventures of the twentieth century were on their way to the sky. One would allow a hundred passengers to cross the ocean every day in just 3 h and 30 min, while the other would take three men to the Moon in about a hundred hours.

On 1 October, Concorde cautiously crossed the sound barrier (Mach 1) without mishap, and the next day, at the Elysée Palace, the new President of the Republic Georges Pompidou confirmed to the president of Aérospatiale Henri Ziegler, accompanied by André Turcat, his determination to pursue the

Concorde program. The flight tests went on and the following month, during flight 102 above the Atlantic, Concorde 001 finally reached Mach 2 for a period of 53 min. This was a speed still largely unknown for such a heavy aircraft, apart from the unfortunate B-58 of the US Air Force.

André Turcat, always attentive to detail, had been able to analyze the fatal accident of the B-58 in Le Bourget, and had drawn invaluable conclusions for piloting his own test flights. The formidable gamble of the Concorde project had been pulled off, a scientific and technical gamble, also requiring considerable courage on the part of the test flight crews and visionary boldness by industrial stakeholders. But the commercial challenge was yet to be met.

Doubtless this victory, both technical and political, owes much to the close relationship between the French people and the world of aviation. In the 1960s, who had forgotten Clément Ader's first motorized flight in Issy-les-Moulineaux in 1897, even though it had been contested? Or indeed, the successive records achieved in the 1900s, and Blériot's channel crossing; or the flying aces of the Great War, such as Guynemer and Roland Garros; or the Aérospatiale epic, connecting Toulouse to Santiago in Chile; or Nungesser and Coli who disappeared in 1927 while trying to cross the Atlantic from east to west in their *Oiseau blanc* (White Bird), then Costes and Bellonte on their *Point d'interrogation?* (Question Mark), achieving the same crossing in 1930; or Saint-Exupéry, pilot and writer, shot onboard his Lightning P38 in 1944, the wreck having been discovered in 2020 offshore Marseille? This public awareness of a historical tradition was as much felt by the powers that be as by industry and its most modest worker, and it undeniably underpins so many of these programs that the French people have reason to be proud of, such as Caravelle, Concorde, and Airbus, to name but the civil aircraft.

From then on and until the scientific flights of 1973, Concorde 001 would continue to undergo tests to define the production aircraft, the first of which took off at the beginning of 1975. But 001 also hosted VIPs, went from Paris to Dakar in 2 h and 35 min, and did a trip to Brazil and Argentina. At the beginning of 1972, the prototype 001 was being prepared for retirement, but the people of Toulouse were by then used to seeing the pure white delta (or more precisely, "gothic") wing and the long-pointed droop nose, like the beak of a strange bird, lowered to give the pilot the visibility required for landing. This will soon be the time where the prototypes 001 in France and 002 in United Kingdom will end test flights to let the scene, from 1976 to 2003, to the remarkable commercial career of the 14 Concorde built. This career was marked by a tragedy, which we must tell and led to the end of this formidable aeronautic adventure.

2.3   In Which a Tragedy Is Endured, But the Business Imperative Wins Out

By 1976, the American and Soviet supersonic projects had been abandoned, whereas the commercial career of Concorde began in January, with a dozen aircraft flown by Air France and British Airways. This part of the story was rather different, unfortunately marred by the United States' refusal to allow access to JFK airport in New York until 1977. There were no accidents during this period, while tens of thousands of admiring passengers were transported, up until the tragic take-off of 25 July 2000 at Charles-de-Gaulle airport in Paris. A tire burst on the Air France F-BTSC carrying a hundred passengers and nine crew members, hits on the runway at take-off a piece of metal lost by the engine of a previous plane. The tire bursts in pieces and one of these hits the wing, causing a kerosene leak, a fire, and the loss of two engines. The plane crashed into a hotel in Gonesse 2 min after take-off, adding four more victims on the ground. While driving my car in the streets of Paris, I heard the radio announce the crash and my heart sank as my mind was suddenly filled with images of the eclipse flight almost 30 years earlier. All Concordes were immediately grounded during the ensuing enquiry, but they regained their airworthiness certificate in August 2001. However, commercial exploitation was halted indefinitely by both Air France and British Airways in 2003. The economic health of the world was no longer what it had been in the 30 prosperous years following the end of World War II. And so ended the dream of a worldwide fleet of more than 1500 supersonic airliners flying in 1990, as suggested in a book[6] published in 1971 and prefaced by Didier Daurat, the founder of the legendary 1930s line to Santiago, Chile, represented by the character Rivière in Saint-Exupery's *Night Flight*. Now passengers from the world over who pass through Roissy Charles-de-Gaulle airport can see Air France's F-BVFF Concorde exposed nose up, as if fiercely ready for take-off.

---

[6] Manel (1971). In this book, prefaced by Didier Daurat, founder of Aérospatiale, the author announced the prospect of a world supersonic fleet of some 1600 supersonic planes by 1990.

3

This Dark Brightness that Falls
from the Stars

3.1   In Which a Navy Officer Becomes
an Astronomer

Jack Eddy was born in the country of the Pawnee Indians, in a small town in
Nebraska (US), in 1931, of a modest family who would be hard put to finance
the studies of their three children with the income they earned from the local
farmers' cooperative, supplemented by his mother's salary as a primary school
teacher. But Jack nevertheless gained a place in the US Naval Academy and
would soon become an officer, particularly interested in what he would have
to know from astronomy in order to take a sight at sea, and much more.
Stationed aboard an aircraft carrier, he took part in the Korean war, but it was
one night in the Atlantic, while on surveillance duty on a destroyer, that his
life was suddenly turned upside-down. A man had just fallen overboard and
Jack requested the order to go back and rescue him from the water. But his
request was turned down, the ship continued on its way, and the drowning
sailor disappeared into the dark night. Immediately, Jack the pacifist, so hand-
some in his officer's white uniform, decided to leave the navy and devote his
life to science, and in particular, to astronomy. Discharged from his military
duties, he entered the university of Colorado, and in 1961 prepared a doctoral
thesis under the supervision of the astronomer Gordon Newkirk, who happened to be one of the leading experts on that part of the Sun which "spills
over" during an eclipse, as Plutarch had noted so long before.

I met Jack and his wife Marjorie a few years later in the Boulder High
Altitude Observatory where I first came into contact with the world of

© The Author(s), under exclusive license to Springer Nature Switzerland AG 2025
P. Léna, S. Koutchmy, *Eclipsed Suns, the Solar Corona and Exoplanets*, Astronomers' Universe,
https://doi.org/10.1007/978-3-031-92199-5_3

aviation, at the foot of the Rocky Mountains, dominating the Great Plains where the Pawnees and their bisons have unfortunately been replaced by motorways and cheap motels. It was Jack, whom we shall encounter again later, who told me his story, and another one too, that of Thomas Edison.

3.2   In Which a Brilliant Inventor Is Engulfed in Feathers during an Eclipse

Born in the middle of the nineteenth century in Ohio, Thomas Edison, fascinated by experimental science since he was a child, would become a brilliant inventor and businessman.[1] Humankind owes him the first phonograph, and also the first electric light, with a filament which becomes incandescent when a current passes through it, transformed into a source of white light. He filed more than a thousand patents and founded the industrial empire that went by the name of General Electric. At 22 h00 on the evening of his funeral—he died in 1931 at the age of 84—President Hoover had all the lights in America turned off in homage. But Thomas Edison also had a brief encounter with astronomy at the age of 31, when the astronomer Samuel Pierpont Langley offered a challenge to the young but already famous inventor, warning him, however, that there would be no financial reward, only that he would do a service to science. Edison would have to invent a highly sensitive instrument to detect, during the next total eclipse of the Sun, the weak infrared light thought to be emitted by the solar halo—the corona. Invisible to our eyes, although some snakes do detect it with theirs, infrared light is the radiation that stimulates in us the sensation of heat when we expose our skin to the Sun or to a wood fire. This infrared light, which we shall discuss further below, was largely unknown in those days.

So Edison set to work and invented the tasimeter, a strange instrument in which the tiny amount of energy received in the form of infrared light would cause the expansion of a piece of rubber by heating it (Fig. 3.1). This in turn would press upon a piece of graphite, whose electrical resistance would thereby change in a way that could be measured. Although he was not a professional astronomer and had little experience in observing the stars, the young Edison, already a star himself in the eyes of the press and public opinion, decided to organize his own eclipse expedition during the summer of 1878, sure that his experimental genius would do much better than all those "mathematicians." He joined a group of genuine astronomers in Rawlins, Wyoming, a small town on the railway line on its way to the Far West, to observe the solar halo

---

[1] Eddy (1979).

Fig. 3.1 Edison's tasimeter. A horn picks up the infrared light coming from the left. It falls on a vertical bar and heats it up, thereby squeezing a piece of graphite whose changing resistance can be measured (Source: American Journal of Science, 1879)

for himself. Having set up his instrument by the door of a small chicken hut to protect it from the wind sweeping across the plain, he carried out a successful night time test on the star Arcturus. Then, in the afternoon of 21 July 1878, he pointed it at the solar halo for the three brief minutes of totality. And lo and behold! The tasimeter did indeed pick up a signal from the Sun.[2]

The rest of the story involves some controversy. The first issue, anecdotal but raised by Edison himself, although he failed to give any details, proved extremely popular with the press over the following few days: Edison claimed to have forgotten that the darkness of a total eclipse disturbs animals, and that the chickens, feeling that night was coming on, had rushed back to the chicken hut in a flurry of feathers, knocking into the crucial instrument and its horrified owner as they went by. The second controversial issue was a question of priority, as is often encountered among scientists, in proportion to their feelings of self-importance. The young Edison, although inexperienced as an astronomer, nevertheless claimed to be the first to measure the faint infrared light from this solar halo, thanks to the extreme sensitivity of his tasimeter (Fig. 3.2). This claim, taken up by the press, was immediately contested. An astronomer from Milan, Professor Luigi Magrini, had already found evidence

---

[2] Eddy (1972).

Fig. 3.2 Numbered points are those where the coronal infrared emission was measured during the eclipse of July 29, 1878 in Wyoming (United States). Their positions are shown against a photograph of the solar corona taken during this same total eclipse of 29 July 1878 in Wyoming (US). (Source: Document US Naval Observatory, 1880)

for this radiation[3] during the brief total eclipse of the Sun on 8 July 1842, visible in the south-east of Europe three years prior to the first photograph (daguerreotype) of the Sun. But Edison had other more important business than digging around in observatory libraries to discover a predecessor who might have overshadowed him!

Whether or not it was the first time in the history of astronomy that an eclipse had been observed in the infrared, this expedition to the Great Plains of the American west already contains all the ingredients that will occupy us throughout the rest of our story: infrared light, the mysterious solar halo, and above all, the eclipse itself. As far as aircraft are concerned, we have already spent some time on them.

3.3   The Invention of an Artificial Eclipse

During the eclipse of June 1973, the villagers of the Sara ethnic group, in the small town of Takamala along the Chari River in the north of the Central African Republic, spoke as follows with three French ethnologists: "Why did you choose us in Takamala? Why did you cause this eclipse here and not at home? The Americans are well on the Moon... You and your Concorde have

---

[3] http://sites.google.com/site/histoireobsparis/table-des-matieres/chapitre-7-arago-directeur-des-observations/eclipse-et-premier-daguerreotype-du-soleil.

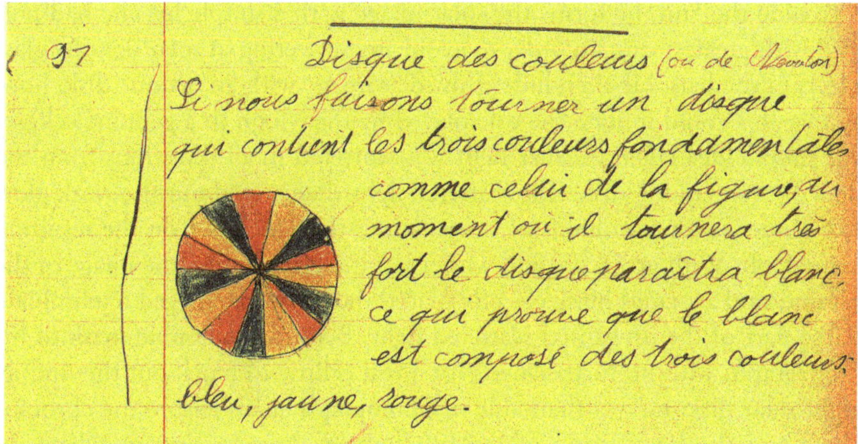

Fig. 3.3 As a child, Bernard Lyot kept a notebook in which he drew and described experiments he carried out himself at the age of 13. Here he describes the experiment with Newton's disk. When rotated quickly, it shows how white light is a synthesis of all the colors of the rainbow. (Source: Club Bernard Lyot)

hidden the sun." Humor, ignorance or irony, whatever; they did not realize the truth in what they were saying as since 1937, astronomers have been able to cause artificial eclipses with an instrument called a "coronograph," invented by the Frenchman Bernard Lyot, which allows to study the parts of the corona which are the closest ones to the solar edge.

Bernard Lyot, born in 1897, was a particularly good student. The scientific notebooks he so carefully drew up himself when, as an adolescent, he was already carrying out his own experiments, could inspire many teachers who would like to give their pupils a taste for science (Fig. 3.3).[4]

When he became an astronomer at the Paris Observatory, he was fascinated by the Sun, which he observed from the world-renowned Pic du Midi Observatory, inaugurated in the Pyrenees in 1882 and famous worldwide for the purity of its mountain sky.[5] Lyot wanted to be able to observe the solar corona every day. At the age of 30, in order to refute the claim by the German physicist Hans Kienle who said in 1929 that[6] "tests prove without doubt that it is impossible to photograph the solar corona during daytime," Lyot invented a subtle instrument which immediately made him famous: the coronagraph.

---

[4] B. Lyot: *Cahier d'expériences* (hand written), kindly communicated to the author by Gérard Lyot, Mireille Hibon-Hartman, and the Bernard Lyot Club.

[5] http://www.imcce.fr/en/observateur/campagnes_obs/phemu03/Promenade/pages5/545.html.

[6] Davoust E., *Le coronographe de Bernard Lyot au pic du Midi*, at http://www.bibnum.education.fr/files/lyot-analyse-43.pdf.

To hide the Sun, he forms the solar image with a simple lens he had polished with greatest care, in order to avoid any scattering of solar light by glass defects in the lens. He then hides this solar image with an opaque disk, hoping that this could imitate the role played by the Moon in a genuine eclipse. This simple idea failed because light was scattered in all directions from the edges of the masking disk. The scattered light then wiped out the weak glow of the inner corona, millions of times fainter than the disk, on the resulting photograph. With great care and with a second lens forming the image of the first one, and a second mask cleverly placed there, Lyot managed to eliminate a large part of the unwanted scattered light. Behind the double mask of his coronagraph, with no further need for a real eclipse to block out the unduly bright solar disk, he was then able to photograph at his leisure the chromosphere and the hot plasma of the bright lower corona, regions where we observe the violent movements of solar matter called protuberances. Nevertheless, the coronal parts, further away from the solar surface and less bright, remain out of reach from the coronograph.

At the Pic du Midi Observatory where he sets up the coronagraph, Lyot produced unprecedented film sequences which showed solar flares as they happened (Fig. 3.4). In 1952, Bernard Lyot went to Khartoum on the Nile in Egypt to observe a total eclipse of the Sun on February 25. He would never see France again, for he died of exhaustion a few weeks later in Cairo while examining his photos. In 1957, the documentary film *Flammes du Soleil* was made by the Paris Observatory in homage to Bernard Lyot. It achieved a huge success when projected in cinemas.[7]

The Lyot coronagraph, which thus creates an artificial eclipse, was soon available in all observatories involved in solar observation. This instrument led to great progress in understanding the lower corona. The years following the Second World War were king years for scientific eclipse expeditions. In this narrow slice of thin solar atmosphere, where the temperature rises suddenly, researchers want to know what is the mechanism that transports upwards so much energy, sufficient to heat the gas to more than a million degrees? Could we still make progress thanks to artificial eclipses? (Fig. 3.5).

Jack Eddy, met earlier in this book at the foot of the Rocky Mountains, having left the United States Navy, tackled the problem in 1960 while preparing his doctorate. Above Wisconsin, at an altitude of 25 km in the stratosphere, a balloon, the Coronascope I, carries an automatic telescope which points at the Sun. The direct light from the latter is masked by an occulting

---

[7]The movie *Flammes du Soleil* (1957) is available from http://videotheque.cnrs.fr/index.php?ulraction=doc&id_doc=1348.

Fig. 3.4 In 1939, Bernard Lyot, inventor of the coronagraph, observing planetary surfaces with his polarimeter at the Pic du Midi Observatory, just before being elected into the French Académie des science. (Courtesy Club Bernard Lyot)

disk located about two meters away from the entrance of the telescope, therefore forming a Lyot coronagraph, but where the occulting disk is external (Fig. 3.6).[8] Without directly measuring the corona during this first flight, Jack shows that at this stratospheric altitude, the brightness of the blue sky is reduced favorably to less than ten billionths of the solar radiance. Four years later, Coronascope II finally obtains, on this principle and outside an eclipse, images of the outer corona. We will meet again the external disk coronagraphs in the second part of this book.

But this great success was beaten to the post, because, a year earlier, a coronagraph, aboard a sounding rocket launched by a US Navy laboratory at an altitude of more than 80 km, obtained the first photograph outside eclipse of the outer corona. Soon after, in the southern Algerian desert, a French team around Professor Jacques Blamont and my friend Roger Bonnet immediately

---

[8] http://www.leif.org/EOS/Eddy/1961_PhD-thesis.pdf.

Fig. 3.5 In 1960, the French astronomer Audouin Dollfus embarks a chronograph, equipped with an external disk, on a stratospheric balloon, to be launched from Aire-sur-Adour (France). Not requiring an eclipse, his observations of the corona, close to the solar surface, confirm the rapid temperature rise in this zone. (Source: A. Dollfus)

renewed this feat by launching another sounding rocket, the French Véronique, also equipped with a coronagraph. This instrument was quite original, since it was equipped with super-polished mirrors in place of classic lenses, in order to reduce the scattered light and operate in the near ultraviolet. Unfortunately, the recovery parachute did not open, and the images could, unfortunately, not be recovered.[9]

As it has since developed, access to space has transformed the study of the Sun and its corona, thanks to the numerous missions that NASA, then Europe, devote to it. The more distant corona, this weak, very extensive luminous envelope, which until then escaped coronagraphs, becomes accessible.

---

[9] Bonnet (2000).

Fig. 3.6 Before taking off in 1964, the balloon is inflated and will carry the Coronascope of High Altitude Observatory (Colorado) into the stratosphere. (Source: with permission of NCAR/HAO)

As the brightness of this faint corona is no longer a million times less than that of the solar disk as in the case of the lower corona, but a billion times fainter. Even from a high mountain, the brilliance of the Earth's atmosphere illuminated by the Sun—the blue of the sky—produces a luminous background on the photographs which hides this faint corona. To observe it, one would have to climb even higher than the Pic du Midi or a stratospheric balloon, and observe for longer than the few minutes of exposure time offered by a rocket.

In fact, the first space station revolving around the Earth, called Skylab and launched by NASA in 1973, was carrying a telescope and observed "the eclipse of the century," as we will see later. Finally, other wavelengths of light have become accessible from space, such as X-rays and far ultraviolet. Their observation revolutionizes the knowledge about the corona, which will be the subject of the second part of this book.

3.4 Going to the Stratosphere to Look for Stars

If I was in this observatory in the Rocky Mountains in 1968, where I became acquainted with Jack the former navy officer, while every evening on the television I could watch in horror the shelling of Vietnam by American bombs with their defoliants, it was because of my own interest as a young astronomer in infrared light.

Right at the beginning of the nineteenth century, in 1800, the British astronomer of German origin William Herschel had produced a rainbow by dispersing the white light of the Sun in a prism, then slowly moved a thermometer from one side of this rainbow to the other, from violet and indigo across to red. For each color, the rise in temperature of the thermometer indicated the energy absorbed from the Sun in that color of light. To his great surprise, Herschel observed that the thermometer reading continued to rise when he moved it beyond the red, thus indicating the presence of radiation that the eye could not perceive. He had just discovered the existence of infrared light. The distribution of this solar light depending on the colors—which we call the "spectrum"—therefore extended beyond the visible spectrum covering from violet to red, over an immense domain which would remain unexplored for a long time, extending from visible red light to radio waves. At the end of the nineteenth century, one would also discover the richness of wavelengths shorter than the violet (ultraviolet, X-rays, and gamma).

To highlight the infrared radiation of distant stars, it became necessary to use energy detectors, then called bolometers, much more sensitive than the Herschel thermometer.

In order to reveal the infrared radiation of remote bodies, it was essential to use much more sensitive thermometers than Herschel, and also to have large telescopes. Even in the middle of the twentieth century, when I was taking my first steps as an astrophysicist, the only infrared light yet detected from celestial sources was from the Sun, the illuminated and hence hot face of the Moon, or indeed the planets Venus and Mars. To complicate matters, even when there is no cloud cover, the water vapor present in our atmosphere only lets through a very little of this kind of light. The only solution was to observe from the tops of high mountains, or from very dry desert regions, so collecting this light without too much loss. Better still, by sending a balloon up to an altitude of 20 or 30 km into the stratosphere, or observing from an aircraft at a height of 10 km, one could considerably increase the chances of a favorable observation, although of course there was a price to pay in terms of the technical complications as compared with an observatory firmly fixed to a mountain top.

In California in 1968, aboard NASA's powerful four-jet aircraft Galileo I, dedicated to other aspects of astronomy as well as eclipse observation, I learnt to fly high to collect the infrared light from the Sun and stars. On the night before the flight, while making the final adjustments in the desert hangar where the plane was kept, standing close to the long silver fuselage which held the instruments that Jack had helped me to design and breathing in the strong smell of oil, I had a marvellous feeling of stepping into unexplored territory. I was terrified at the thought of the failure that would result from the slightest error in the alignment of my telescope or the adjustment of our bolometers,[10] now cooled to a temperature of 269° below zero to make them more sensitive (Figs. 3.7 and 3.8).

Five years later, it was with great sadness that I learnt on 12 April 1973, while in Toulouse preparing the Concorde flight, that Galileo I had been destroyed on its return from a science flight with 12 men aboard, after a terrible collision with a small navy aircraft, just before coming down on the landing strip where we had spent so many nights. Its successor Galileo II had a less tragic fate, without victims, a few years later, catching fire on landing after a science flight. Science is not always plain sailing.

It was 1969 and France was still feeling the consequences of the civil unrest of May 1968. The Paris Observatory and the Sorbonne, which I had just entered, were as shaken up as the rest of France, or even more so, with

Fig. 3.7 From 1974, the Paris Observatory flew a telescope aboard the Caravelle 116 stationed at the Brétigny test flight center in Essonne, France, photographed here as it approaches the center. The telescope, designed to observe stellar infrared radiation, is set up in the position of the port side emergency exit. This Caravelle is now exposed at the aerodrome in Montélimar, France. (Source: P.Léna/Centre d'Esssais en Vol)

---

[10] The bolometer built by the physicist Frank Low (1933–2009) was so sensitive to detect infrared radiations, that it soon equipped observatories all across the world.

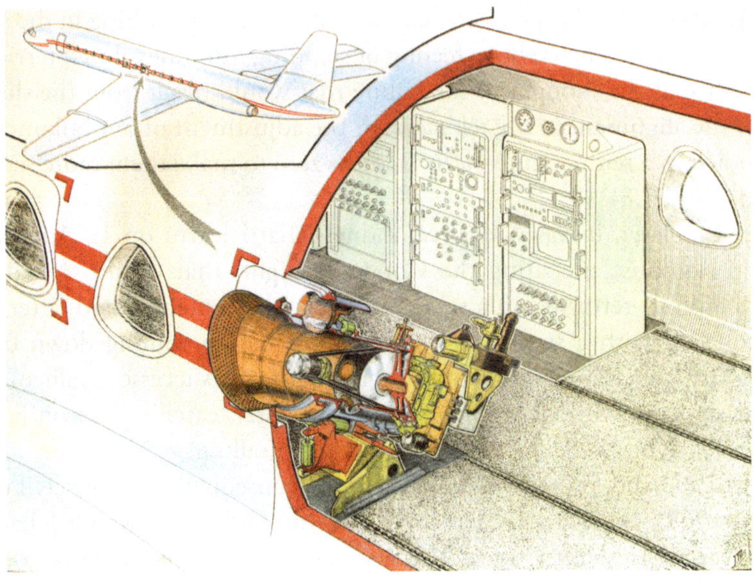

Fig. 3.8 Artist's view of the telescope set up in the Caravelle 116 aircraft by the Paris Observatory, before being transferred to NASA's Galileo plane. Between 1972 and 1979, with the PhD students Daniel Rouan, Marie de Muizon and others, this instrument would make many astronomical observations of infrared radiation in both the northern and southern hemispheres (Source: Drawing by G.Mir/CNRS)

astronomers dreaming of putting the world to rights, if not the whole universe. So why not go on exploring the infrared sky with a plane? It was to be a Caravelle reserved for test flights, carrying the number 116, a marvellous twin-jet already mentioned in this story, that would host our telescope, specially built to fly on this plane, year after year, as high as possible, and at so many different latitudes (Fig. 3.9). We often felt like Fabien in the famous book *Night Flight*, when he emerges into a star field[11]: "Their pale magnets attracted him. He had struggled so long for a glimpse of light that now he would not have let even the faintest get away from him. [...] And thus he rose toward these fields of light. [...] He wandered amid stars gathered with the density of a treasure, in a world where nothing else, absolutely nothing else other than he, Fabien, and his comrade, was living".

A smiling young man, Claude Nicollier, barely 30 years old, passionately interested in astronomy and aviation, had just joined our group from Switzerland.[12] Captain in the Swiss air force, he roamed the sky and the mountains at the speed of sound. After his PhD in astronomy, he became an

---

[11] Saint-Exupery (1974) in Chap. XVI.

[12] http://fr.wikipedia.org/wiki/Claude_Nicollier.

Fig. 3.9 The whole team from the Paris Observatory aboard NASA's Galileo I aircraft in 1976, with two of their European partners, gathered around the telescope (below the blue tube in the background). Many flights were carried out over California to explore the infrared sky. From left to right: top, John Beckman (UK), Pierre Léna, Jim Lesurf (UK); bottom, Guy Vanhabost (F), Jan Wijnbergen (ND), Pierre Gigan (F). (Source: P.L.)

airline pilot for Swissair, but he could not resist the attraction of astronomy and soon returned to it. Europe then selected him as an astronaut aboard the Space Shuttle. Our telescope and its observation program were chosen to carry out an in-flight training program for astronauts, prepared on the Caravelle and carried out aboard NASA's Galileo II plane in the USA. With exceptional kindness, Claude taught me the patience and absolute rigor required for his job, qualities I seriously lack. A decade later, he went on three extraordinary astronomy missions into space aboard the Space Shuttle, including one to repair the Hubble Space Telescope.

In our work on Galileo I and II and on the Caravelle, we built up a strong and friendly team of researchers, students, technicians, and engineers.[13] We had learnt how to fly toward the stars, our financial backers trusted us, and the infrared sky was opening up to our efforts and those of other astronomers around the world, with several discoveries on the horizon. So enter Concord 001!

---

[13] Rouan (2012).

4

The Run Up to the Longest Total Eclipse in History

The site in Blagnac near Toulouse is part of an aviation legend. It was here that Didier Daurat and Pierre-Georges Latécoère created the Aéropostale line to South America, made so famous by Henri Guillaumet, Antoine de Saint-Exupéry, and Jean Mermoz. It was from here that they took off from a short runway, now almost covered with buildings, with their precious cargo of mail, toward the south and the razzias of the African coast. And in Dakar, it was with great emotion that I came to contemplate the stone monument[1] to Mermoz and his companions who fell from the sky, now standing at the site of the landing strip where they came down before setting off across the South Atlantic in their Laté-28 seaplane, at the intersection of the *rue de la Pyrotechnie* and *avenue Cheikh Anta Diop*. And in this year of 1972, it was also from this site in Toulouse that the prototype 001 of Concorde would take-off, week after week, up until the time a few months later when all the test flights had been successfully completed under André Turcat's leadership.

4.1 A Crucial Lunch Meeting

That same spring of 1972, I made a phone call from my laboratory in Meudon, requesting a meeting with André Turcat, engineer and pilot, who was in charge of the test flights, although I did not mention my reasons or discuss them with anyone. The man was already of considerable renown for his aviation records and highly esteemed, but he kindly accepted the request from

---

[1] http://www.aerosteles.net/fiche.php?code-dakar-sacrecoeur-mermoz.

P. Léna, S. Koutchmy, *Eclipsed Suns, the Solar Corona and Exoplanets*, Astronomers' Universe, https://doi.org/10.1007/978-3-031-92199-5_4

this young and unknown astronomer, and very courteously invited me to lunch with him at the restaurant La Fontaine at the airport when I disembarked from the Caravelle used for my Air Inter flight from Paris. Perhaps this was a good omen? On the paper napkin covering the table where we lunched looking out over the landing strip, I quickly outlined my dream, sketching a map of West Africa and explaining that the Moon's shadow would be moving at close to the maximum speeds that could be reached by Concorde. I also detailed my modest experience of aeronautics and concluded by describing the extraordinary record which lay within our reach: the Sun would remain totally hidden to us for an incredible 80 min, more than ten times the longest time that would ever be possible from a fixed observation point on the Earth's surface. André Turcat gives the following account of this first meeting[2]:

A relatively well-ordered shock of black hair, with likewise black and shining eyes behind his spectacles, this astronomer had just sat down at the table before me and began to present his project at some length. 'All that was needed' was a supersonic aircraft, admittedly a little expensive, with a few holes made in the roof of the cabin, a few portholes made out of a special material, an electricity supply for his astronomical observation equipment, and extremely accurately programed flights over Mauritania and Chad. Concerning the storms over the intertropical convergence zone, this young astronomer had little time for them. His final summary was brief: 'So what about it?' Yes, indeed! I was won over and my enthusiasm matched in every way that of this young professor. But my mind was already occupied with all the problems we would have to face.

When the young astronomer flew back to Paris that evening, a plan of action had already been sketched out, for the days were counted. It was May and the eclipse would occur in June of the following year. My laboratory was very busy with many other projects and I wondered how this one would be received by my colleagues. Who would finance it? Where would I find the much-needed support and cooperation? When I got home, I was suddenly terrified by my own audacity and the risk of failure after having dared to involve such a prestigious aircraft. My wife encouraged me and I took the little globe I used to introduce our first two children to the fact that the Earth was round, trying to reassure myself by considering at length the possible trajectories (Fig. 4.1).

But before solving all these problems and resuming contact with the research unit in Toulouse that André Turcat had immediately put in the picture, there was one key question I had to answer. Mobilizing Concorde and

---

[2] Turcat (2013). A less detailed account is in Turcat (2020).

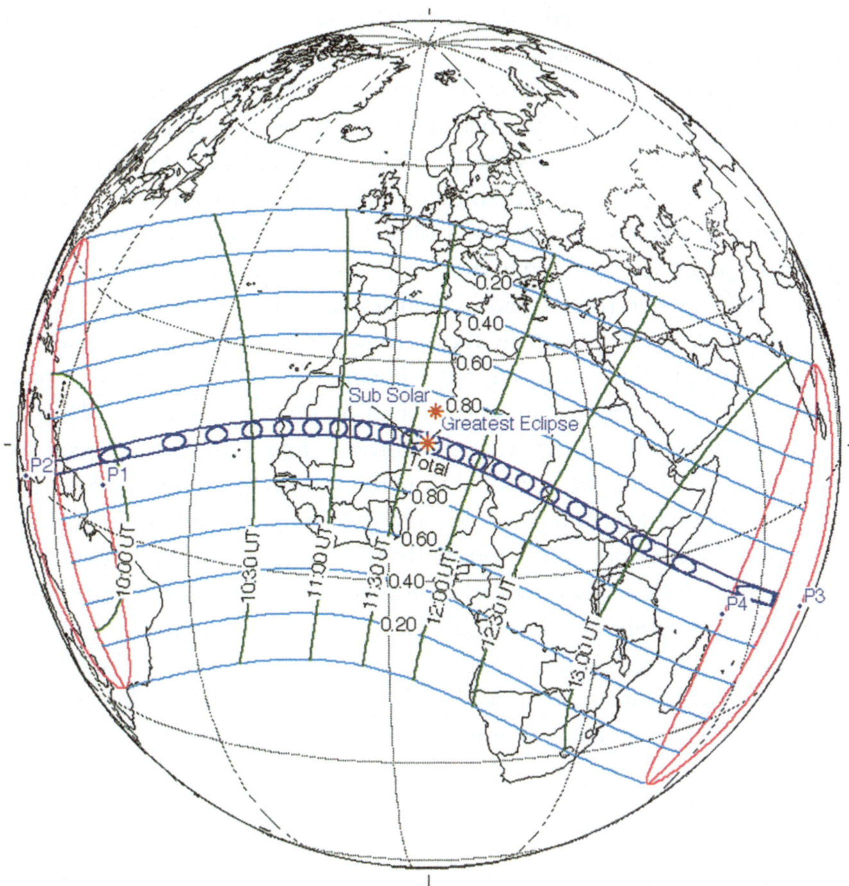

Fig. 4.1 Dreams inspired by a small globe. The narrow path of the Moon's umbra from P1 to P4 (dark blue) and penumbra (area covered by light blue lines) as predicted by celestial mechanics (Source: Eclipse predictions, Fred Espenak, NASA/ GSFC)

chasing after the shadow of the Moon was all very well, but to achieve what exactly? What scientific problem was I going to tackle? What unresolved question would I attempt to answer by this novel astronomical observation, made possible by the exceptional duration of totality observed from the Earth's stratosphere?

4.2   Corona, Dust, and Infrared Light

"Interplanetary space", the space that separates us from the Sun, is not completely empty of matter. The Sun constantly ejects hydrogen and helium in the form of electrically charged ions, while tiny grains of dust remain in this space, fossil, and fragmented witnesses of the Solar system a few billion years ago, or residues of comets evaporating when approaching the Sun. Electrons and protons, issued from dissociated atoms of hydrogen and helium, as well as the dust particles create this solar halo first noticed by Plutarch during a total eclipse. Since it has become a subject of scientific investigation and detailed observation, this rather faint envelope has become known as the solar corona, with its two components K and F. Briefly described in our introductory chapter, it will be analyzed in more details in the second part of this book.

In the 1970s, when the astronomy of infrared light was coming into being, the cosmic dust, omnipresent in the universe, was beginning to arouse the interest of astrophysicists. Indeed, our telescope aboard the Caravelle aircraft had been set up to study them, and we had become specialized in the main techniques. While the existence of the coronal dust particles had been suspected, or even established, for some time then, their low temperature, hence weak infrared radiation, and I have already mentioned the difficulties involved in detecting these infrared radiations with "thermometers" of high sensitivity .

During an eclipse some time earlier, Robert MacQueen, an American colleague I had met while working with Jack Eddy on flight preparations for the NASA four-engine jet Galileo I, had in fact been trying to repeat the observation of the solar corona carried out by Thomas Edison. Guided by some reasonable hypothesis, he had come up with a rather interesting result. Not far from the surface, the Sun seemed to be surrounded by rings or shells of tiny dust grains—silicates, in fact—which concentrate in space before evaporating under the effects of the intense radiation from our star. Therefore, a zone "forbidden to dust" could exist from the Sun surface to about four solar radii.

For their part, physicists are calculating how a tiny rocky grain of interplanetary material behaves—a fraction of a thousandth of a millimeter—when it is subjected to the gravitation of the Sun, which attracts it, while receiving its light which repels it,[3] heats it and, if necessary, vaporizes it. In addition, various types of crystalline grains, well known from the collection of shooting star dust in our atmosphere or even in space, behave differently, depending on whether they are iron-based olivine or magnesium, to name just two minerals from the silicate family. The question is therefore not that of the temporary presence of such grains in the immediate vicinity of the Sun, before they evaporate, heated to more than 1500 °C. But can these grains concentrate enough to form rings or shells?

---

[3] The pressure which light exerts on an illuminated dust grain, called "radiation pressure" is detailed in Chap. 8.

These heated grains, about a thousand of millimeter in size, would emit some infrared light. They may be concentrated in the plane of the orbits of all the planets, the so-called ecliptic plane, forming rings, or perhaps surrounding the Sun entirely in the form of shells. It was already understood without formal demonstration that, one way or another, over millions of years, these grains were capable of sticking together to form ever larger ones, and eventually exceptionally large bodies like our own planet. We knew that, between the planets Mars and Jupiter, asteroids collided during the long history of the solar system, the collision producing dust grains. We also knew that comets, when approaching the Sun, were evaporating in a long and dusty tail.

The grains certainly deserved investigation in order to explain their presence, if confirmed, so close to the Sun. Would we thus be able to improve the result obtained by our colleague by exploiting the advantages of a flight that could provide more than a whole hour's worth of corona observations, rather than just a few brief minutes, without being dazzled by the intense radiation from the solar surface? Better still, flying at an altitude of 17,000 m in the stratosphere, Concorde would be well above the water vapor which is present in the lower Earth atmosphere and which significantly attenuates infrared transmission. We should be able to confirm the existence of rings and perhaps even determine their chemical composition. I discussed all this with my colleagues and made a few preliminary calculations, whereupon we agreed that it was feasible. Our subject would be the F-corona, that is, the dust in the corona. Shortly later, during the autumn, I went back to the USA and described the project to my colleague Donald Hall, a brilliant As early as 1971, 28-year-old Australian who had also specialized in solar infrared radiation. Enthusiastic, he accepted to join our project and got the agreement from the Kitt Peak Observatory in Arizona where he was working.

I now had a firm scientific motive for undertaking this adventure. It remained only to make a formal announcement, set out the details and feasibility, and find funding. And time was short, too short, for there were three irons to get in the fire: convincing the supervisory authorities, beginning the close collaboration with the research unit in Toulouse in order to determine a possible refurbishment of the plane, and building a completely new instrument suited to the kind of infrared observations we were hoping to make.

4.3   In Which a Future Research Minister Gives his Support

I put together a small team thanks to the immediate support of close and trusted colleagues, thus reducing the very real risk of overload that my initiative might impose on the programs already undertaken by the laboratory. From then on, Charles Darpentigny, Alain Soufflot, Jacqueline Mondellini, and Yves Viala were constantly by my side. Despite the extra workload, I continued to give my physics courses at the university, without daring to tell my students about this slightly crazy project that was in fact occupying all my thoughts. Naturally, the first thing I did was to get the agreement from the appropriate research authorities, without which I could do nothing. The president of the Paris Observatory was then Raymond Michard,[4] and as luck would have it, this eminent solar specialist, somewhat shaken by the events of May 1968, had also carried out an observation of a solar eclipse in Indochina 20 years earlier, aboard a Nord 2501 military aircraft. Needless to say, he gave us his full support, provided that it was not going to cost too much. The necessary financing could only come from higher up. Now, our laboratory also depended on the French national research organization (*Centre national de la recherche scientifique* CNRS) and a completely new institute had just been set up in response to the upheavals of May 1968. Those in charge were astronomers, fully aware of both the interest and the risks of this slightly mad undertaking. But they gave their support.

I also went to see Pierre Aigrain, then called the Delegate General for Scientific and Technical Research, since there had never yet been a genuine minister for scientific research under the Fifth Republic in France. As early as 1971, the President of France Georges Pompidou had flown on Concorde 001, with André Turcat at the controls, to meet the then President of the United States Richard Nixon (invested in 1969) in the Azores. Regarding Pierre Aigrain, a future minister himself, he was the Prime Minister's acting Delegate General for this domain. He was a brilliant physicist, literally abounding in energy, who had taken me into his laboratory at the *Ecole normal supérieure* some 15 years earlier, helping me to develop a first infrared-sensitive "thermometer" while I was still a student. I met him on a beautiful September morning and explained the project and the science we expected from it. Drawing as always on his pipe, it only took him a few minutes to pledge his support and some of the funds necessary for the construction of

---

[4] http://www.bureau-des-longitudes.fr/membrescorrespondants/michard.htm.

our instrument. The CNRS and the national center for space studies (*Centre national d'études spatiales* CNES), the two large public bodies that financed our laboratory, also approved the underlying idea of the project, despite the vagueness pervading certain aspects of it. The Minister of Finance even agreed to exonerate the whole operation from the imposition of tax.

For his part, exercising his moral and technical authority, André Turcat took similar steps in Toulouse, approaching the managing director of Aérospatiale, Henri Ziegler. The latter, an ex-student of the *Ecole polytechnique* who had not forgotten his love for science, was Commander in Chief of the French Forces of the Interior (*Forces Françaises de l'Intérieur* FFI) in the French Resistance, and also a former test pilot. At this time, he was also one of the founders of the Airbus program, with launch of the Airbus-300 airliner. He viewed the eclipse project favorably, and later on, having received a full technical report, he accepted that Aérospatiale should cover the costs of the transformations required to refurbish the plane, as well as the cost of the eclipse flight over Africa and the necessary rehearsal flights.

In autumn 1972, I was told that we could begin work, but that no firm decision about the eclipse flight would be taken before February 1973, when we would be in a position to make a final assessment of the preliminary studies and costs. We only had four short months to prepare for the big day.

4.4 How Many Holes Should Be Made in Concorde's Fuselage?

Rather soon, the idea of flying such a big plane for our group alone began to look somewhat unreasonable, not just to me, but also to the authorities. It was clear that we should make this an opportunity for other astronomers, in fact, as many as the aircraft could safely carry. But how should they be chosen? Another French team stood out. At 32, Serge Koutchmy was already an accomplished eclipse hunter. He would eventually become one of the best specialists in the world for observations of the solar corona. Astronomer at the Paris Institute of Astrophysics (IAP), he was thus preparing for the eclipse of the century by planning two expeditions, one in Mauritania and one in Chad. These would prove extremely successful, allowing him to obtain first rate color photographs of the corona (see Fig. 1.1). He was thus unavailable for the Concorde flight, but put forward a carefully considered corona observation program based on techniques he knew well, which were thus less adventurous than our own infrared program.

With the resources of the Institut d'Astrophysique de Paris then directed by Jean-Claude Pecker, to whom I owe a good part of what I then knew about the Sun, Serge designed an instrument complementary to his projects on the ground, and delegated one of his colleagues, Jean Bégot, to implement it on board the plane and fly with us. In the United Kingdom, demonstrating the close Franco-British collaboration around this aircraft, another Concorde prototype, the 002 had been built, in particular to develop the engines built by Rolls Royce. However, it would only make its first flight on 10 January 1973, 13 months after the maiden flight of Concorde 001. The British astronomer John Beckman, himself interested in solar infrared radiation, had recognized Concorde's potential for observing the eclipse and had approached the British Aircraft Corporation (BAC) which had built the 002 prototype. The response was negative. The company was concerned about upsetting its schedule, and in particular about creating a new porthole in a plane that had not yet even started its test flights. Although I had been quite unaware of these goings on, I already knew of John's scientific work and greatly appreciated his experimental abilities. We thus offered him a place aboard 001, and he accepted. His scientific program would study the narrow zone immediately around the rim of the solar disk, a region known as the chromosphere. The name derives from the Greek word *chromos*, meaning color. Indeed, the light emitted by the hydrogen present in this very thin gaseous layer is a beautiful red color, very briefly visible during an eclipse at the instant when the lunar disk, just as it masks the Sun, allows a glimpse of this thin band of light. But what lasts on the ground only for a brief instant would last ten times longer with the help of Concorde, slowing as it does the vast sweep of the Moon's shadow. John, who hoped to study this poorly understood zone, would thus have much longer to measure its radiation. He, too, would benefit from the stratospheric flight to gain access to an infrared radiation that is practically impossible to observe from the Earth's surface. Hence, in 1966, prior to John's project, an expedition by a few young American colleagues who would subsequently become highly reputed astronomers had been to the region of Arequipa in Peru, at an altitude of 4500 m, to try to detect this radiation from the chromosphere, although with somewhat limited success.[5]

So that was already three holes that would have to be made in the fuselage. They would be made in the ceiling of the cabin since, during the eclipse, at the summer solstice and practically on the Tropic of Cancer, the Sun would be close to the zenith, that is, vertically overhead. We wanted them to be the maximum authorized size allowed by safety considerations. Indeed, the

---

[5] Noyes et al. (1968).

special transparent windows to be placed in these portholes would have to resist the pressure difference between the cabin and the outside, but also the high temperature—around 100 °C—reached by the outer surface of the fuselage when in supersonic flight, simply due to friction with the surrounding air. If the porthole were to explode in flight, although it would not be a major disaster for the plane and its passengers, it would nevertheless mean an emergency landing and a sudden end to all our observations. The Aérospatiale research unit generously authorized a further hole to be made (Figs. 4.2 and 4.3), so we were able to invite the American astronomer Donald Liebenberg aboard, another specialist in the study of the solar corona who, with his flight

Fig. 4.2 In the spring of 1973, in the Aérospatiale hangar in Toulouse, four extra portholes were made along the axis of the Concorde 001 cabin (arrows). It would be through these openings that the four teams of astronomers would observe the solar corona on the day of the eclipse. The fifth experiment would use an already existing side window. To check how well these windows would stand the pressure difference generated by a stratospheric flight, the plane was literally filled up with compressed air in the hangar. (Source: P.L./Aérospatiale)

Fig. 4.3 The porthole cover in the form of a four-leaved clover, seen from inside the cabin. Through this the instrument from the Paris Observatory would be able to view the Sun. It is still visible on the prototype Concorde 001 exposed at the aerospace museum in Le Bourget, France (Source: P.L./Aérospatiale)

over Samoa in 1965, had already written a new chapter in the history of airborne eclipse observation. He belonged to the long line of pioneers that began with Mahieu in 1912, then Stevens in 1937, Michard in 1955, and several others after them.

In 1963, spelling the end of what was undoubtedly the most frightening period of the Cold War between the Soviet Union and the USA, an international treaty was signed which prohibited nuclear tests in the atmosphere. To monitor explosions carried out by the other side by flying at high altitude and outside the borders of the Soviet Union, the US Air Force had equipped three subsonic four-jet aircraft (NC-135) with suitable detection systems, and these were suddenly freed up as a consequence of the treaty. One of these was subsequently dedicated to the scientific study of solar eclipses.[6] The young Donald Liebenberg took part in the expedition which went to Pago Pago in Samoa in 1965 and obtained excellent in-flight images of the corona during almost three and a half minutes of totality. Just prior to this, the plane had come upon some annoying high-altitude clouds and the pilot had had to use a rather hurried emergency procedure, normally only allowed in time of war, to climb higher without delay.[7] Donald had then taken part in all the NC-135 missions in Brazil, Texas, and Canada, he was in the line of pioneers such as Mahieu in 1912, Stevens in 1937, Michard in 1965, and a few others. It was

---

[6] Dolci (1997).

[7] Mulkin (1981), https://sgp.fas.org.

thus natural enough that we should invite him to fly aboard our 001 so that his instrument could benefit from the fourth porthole. Quite impolitely, we could only provide him with an uncomfortable folding chair, rather than an ordinary seat, so as not to overload the plane at take-off.

To complete the team, a fifth partner, Paul Wraight from Scotland, will join. He does not need an extra aperture in the fuselage, since he would not aim at the eclipsed Sun, but study the effect produced by the sudden darkness in which the eclipse immerses the oxygen atoms of the Earth atmosphere. To achieve this, it will be enough to aim slightly above the horizontal through a normal passenger window of the plane whose glass will have simply been replaced by a transparent and well-polished crystal.

4.5   Some Very Demanding Astronomers

In close relation with the astronomers and engineers who would accompany us, the Aérospatiale research unit was working on the problem of the portholes, the electricity supply for the instrumentation, the authorized mass of the aircraft, and even the number of seats possible in the remaining free space, which would determine how many people would be lucky enough to fly with us. We could already tell that seats would come at a price and I was careful not to make promises to anyone.

But the flight? Would it really come off? Let us hear once again what André Turcat had to say:

I immediately entrusted this preparation to Henri Perrier, the chief navigation engineer, faithful since the first flight. For me there remained above all the preliminary question: the decision about whether it was wise to land at Fort Lamy (today N'Djamena) at around 2 p.m. on 30 June, a bad time for meteorological reasons. What would be the fuel reserve upon arrival if I had to wait for a storm to pass? And within the framework of all our work aimed at test flights, we would be able to prepare and include our eclipse, each of us concerned with the need to be thorough and with the desire to be conclusive and not to miss out on this extraordinary opportunity. But there would be no point in getting enthusiastic about something that turned out to be impossible, would there?

And to begin with, what should be our starting point? Dakar? Perrier had quickly ruled that out. The runway was not very long, and well cleared of obstacles, but above all, it would be hot at the date and time of take-off, and this would reduce the maximal allowed mass at take-off, that is, the amount of fuel. Furthermore, the weather conditions could be poor in the harmattan or dust haze.

I suggested Sal, an island in Cape Verde, where we had already made a stopover on the way to South America. It was a good runway with very regular weather conditions, always north of the intertropical convergence zone, but it would be difficult to lodge any large number of people, for there was no water on the island, simple bed and board but no hotel, and no regular air link.

Perrier came up with Las Palmas in the Canaries. There were none of the disadvantages of Sal. The weather was less reliable but never bad, the temperature was a little higher but never above 26 °C. However, Las Palmas was a little further from the rendezvous and would require take-off in the wrong direction, into the trade winds. Perrier showed me that it was rather an advantage, avoiding a costly U-turn during the climb. We discussed all this. A more precise calculation of the details would decide, so we sent all the facts to the computing unit. The result came back the next day: 2 or 3 tons more for Las Palmas. Perrier was right, as always. And provided that there were no modifications to the details of the flight path by the astronomers, I would have 10 tons of kerosene upon arrival in Fort Lamy, 40 min' wait, and the right to a missed approach. It was reasonable, but a storm can last an hour, and imagine an approach in very poor weather conditions if one is running out of fuel. Consulting a booklet on the local weather, I was pleased to learn that, in this period at the beginning of the rainy season, storms rarely occur before 4 or 5 p.m. local time. Further, a phone call to colleagues at the company UTA, familiar with the area, confirmed this information and told me more, invaluable this time: it was exceptional for Fort Lamy and Kano in Nigeria to be blocked at the same time. Kano had a good runway where I had once landed a military Dakota aircraft. That would leave me the possibility of a rerouting decision up until the moment of descent.

Astronomy also imposed constraints on the Concorde flight path, although of another kind. The geometrical features of the eclipse raised no difficulties. The exact position of the center of the Moon's shadow on the surface of the Earth, or at around the altitude of 17,000 m at which we would fly, the exact size of the shadow which strictly specified the region in which the Sun would be totally hidden, and the speed at which it would move relative to the ground were all perfectly computed and freely available in tables published for each future eclipse. In my notebook of the time, I find my first estimate of the maximum duration of totality, based on a possible speed of the aircraft and in the absence of wind: 83 min and 58 s! But for this, the pilot must first have mapped out the ideal flight path and inserted the details into the onboard computer, informed by the two onboard inertial guidance systems. Their gyroscopes hold a fixed position relative to the stars and can be used to determine the geographical location on the map at any time, hence also the distance covered since the point of departure of the flight. Any inaccuracy in these systems would lead to an error in the rendezvous, any breakdown to a missed rendezvous, whence the aircraft had two, working independently (Fig. 4.4).

In the USA, Donald Liebenberg had all the tools needed to carry out the necessary calculations for an observation from a rapidly moving aircraft. He suggested a flight path made up of three arcs of a great circle. The advantage here was that each of these three "straight lines" plotted on the terrestrial sphere would allow the plane to fly without bends, simplifying the pilot's task and also the accurate pointing of our instruments. In the end, from Donald's calculation, we chose a path made up of a single arc of a great circle, more favorable for carrying out observations because there was no bend in it at all. It would touch the edge of the shadow at the instant known as second contact, when the solar disk suddenly becomes invisible, cut directly across it to reach its southern edge, then as the shadow overtook us, leave by its northern edge, after which we would come to land. If there were no opposing wind, this path would give us 80 min of darkness. Demanding as ever, we also

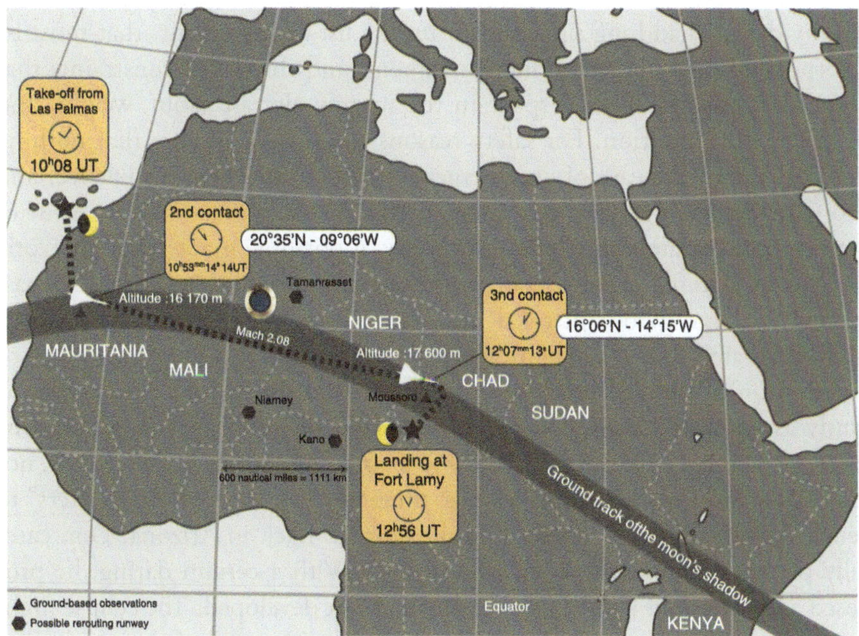

Fig. 4.4 The flight plan for 30 June, designed to chase the Moon's shadow across the African continent. Total darkening of the solar disk occurs between the second and third contacts, those instants when the lunar limb and the solar limb touch one another. First and fourth contacts refer to the beginning and the end of the eclipse phenomenon. (Source: After G. Zimmerman)

required this rendezvous with the shadow during second contact to occur to within 15 s of a fixed time, at a fixed geographical point over Mauritania. We argued for maximal altitude. That was no problem because Concorde flies higher when the atmosphere is colder. But close to the equator, the stratosphere is colder than at medium latitudes and we could count on a favorable average altitude of 17,000 m. We also wished to slightly exceed Mach 2 in order to follow the shadow for as long as possible. The authorized speed limit for the 001 prototype was Mach 2.05. However, if the atmosphere is colder, the speed of sound (which depends on the air temperature) is lower there, and Mach 2.05 would give us a lower speed relative to the ground, hence also relative to the shadow, thus reducing the possible duration of totality.

All this was communicated to André Turcat so that he could prepare the flight schedule down to the slightest detail. On 2 February, the decision was taken by the authorities: we would fly!

4.6   Hands-on

On my desk, I had long since drawn up the list of ingredients that must be incorporated into our observing instrument, and the list of constraints that had to be satisfied: limited space in the narrow aircraft cabin, weight, and electricity consumption. For safety reasons, we had to be sure that nothing would move in the case of an emergency landing or severe turbulence, and also that there was no risk of a short-circuit that might start a fire. A list of somewhat disparate items[8] was soon ready: a porthole made from an exotic and costly material called Irtran, which would allow infrared light to pass through; a large gyro-controlled movable mirror mounted on rails, which could accurately reflect the light whatever the residual motion of the plane or the position of the Sun relative to the plane (and this would change significantly as we were flying to the east); to make an image of the Sun and its corona, a telescope that would reduce to a large and light aluminum dish, not too costly; and finally, the key feature, three ultrasensitive "thermometers" to detect the infrared light emitted by the corona. Back in Arizona, Don carefully prepared two of these "thermometers." With a certain daring, he proposed a new kind of instrument that he had just developed. The eclipse flight would be its first application, but later this device, made from two rather exotic chemical elements, indium and antimony, would be widely used in

---

[8] The French would call it a list *à la Prévert*, with reference to Jacques Prévert's famous poem *Inventaire*, which lists all sorts of items with no apparent relation between them.

many observatories around the world. Regarding our third "thermometer" (a bolometer), I owe this to the young French scientist Noël Coron and his skillful technical team. A brilliant experimental physicist but sometimes tough to deal with, he imposed drastic conditions on the loan of his equipment, but as an amateur pilot himself, he could hardly have done otherwise and accepted gracefully. These three detectors of infrared radiation could only work correctly at unbelievably low temperatures. To keep them cool in flight, we had no choice but to use liquid nitrogen and helium. These precious cooling fluids would have to be supplied in Toulouse and taken to Africa in large Thermos flasks. If by bad luck they were to evaporate too quickly, our measurements would become impossible, and so also would John's, for his instrument would also depend on it. Finally, we would need to add a touch of electronic circuitry to our list to get the whole thing working; a pump to lower the temperature of our "thermometers"; two large recording devices, one with pen and ink and the other with magnetic tapes, which would store our results and which were generously loaned to us.

All we had to do was put the whole thing together. My two engineering accomplices at the CNRS, Charles Darpentigny and Alain Soufflot, worked marvels, always calmly and in good humor, trying for my own impatience, for the deadlines were coming up fast. The first had a long experience in aeronautics, having started his career in Nord Aviation and then designed our telescope on the Caravelle (see Fig. 3.8). The second had already been entrusted with the task of building a French instrument for solar exploration which would be sent into space by NASA in 1975, aboard the Orbiting Solar Observatory (OSO-8). Drawing, cutting and milling, mounting and aligning, plugging in wadges of multicolored electrical wires, everything fell into place with a healthy mix of do-it-yourself and professionalism, while keeping an eye on a budget which had little elasticity and on the weight which could not exceed 400 kg. By the month of April, we were able to send our strange machine to Toulouse and fit it aboard the 001 to see if it would work correctly (Fig. 4.5). In the huge hangar that housed the aircraft, there was a persistent smell of oil which, like one of Proust's celebrated madeleines, took me straight back to those nights in California 5 years earlier, when I had stood next to NASA's four-jet Galileo I aircraft, my first encounter with the world of aviation. One afternoon in Toulouse, we were alone in the plane fixing up the circuitry when a strong smell of burning suddenly came to our notice, causing general alarm. We were not proud of ourselves, until the plane's technical team noticed that the short circuit had occurred in a quite different location and thereby cleared us of responsibility for any false manipulation. Work went on, and in the meantime our British and American colleagues were each

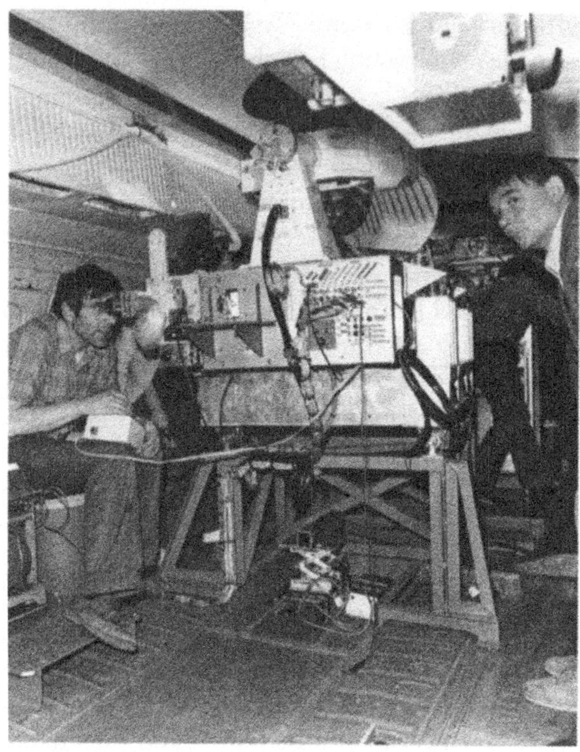

Fig. 4.5 The instrument built at the Paris Observatory and the prototype department of the CNRS is placed vertically below the porthole, where is seen the large gyro-stabilized mirror, lent by NASA and checked by engineer Douchane Stevanovitch in Meudon. The author (P.L., left) is controlling the optical alignments. The engineer Charles Darpentigny (right) designed the instrument, taking great care to respect aeronautic safety standards. (Source: P.L.)

setting up their equipment. In fact, a fifth partner, Paul Wraight, had come from Scotland to join our conspiracy. This physicist did not require any further hole to be made in the cabin, for he was not planning to observe the eclipsed Sun, but rather the effect produced by sudden darkness on oxygen atoms in our terrestrial atmosphere. To do this, he needed only to aim slightly above the horizontal, through a normal porthole whose pane had simply been replaced by a transparent and more carefully polished crystal window.

4.7   Atmosphere and More Atmosphere

Week by week, in the hangar in Toulouse, the team's excitement was mounting, and all the while the deadline was moving closer. Fortunately, the plane was not required to fly too often. But why was the 001 prototype, originally programed to end its test flight career at the end of December 1972, allowed to go on for a further full semester? Was it for the exclusive but costly benefit of a handful of lucky astronomers? Not exactly.

Apart from the visible light perceived by the retina in our eyes and the infrared radiation picked up by our skin cells, the Sun also emits ultraviolet (UV) light which, in small doses, tans the skin of bathers and mountain dwellers, but at the risk of sometimes dangerous melanomas. But in larger amounts, this UV radiation could wipe out all forms of life on Earth. Fortunately, though, above an altitude of 15 km, the stratosphere contains a gas called ozone, derived from the usual dioxygen molecules O_2. Molecules of ozone each contain three oxygen atoms rather than just two. This blocks out most of the solar UV and thus plays the role of an invaluable protective filter. But how fragile is it? Given its crucial importance, the question had long been raised by chemists, but it was not an easy one to answer. Indeed, it was often enough for the molecules to be present in very small amounts in order to upset the complex chemical equilibrium which had maintained this atmospheric filter in place over the centuries. Some of these molecules had been identified in quite an unexpected way. In 1942, in Belgium under German occupation, professor Marcel Nicolet had been ordered to interrupt his meteorological studies. He subsequently took up an interest in the chemical equilibrium of the stratosphere and discovered the role played by certain molecules, in particular, oxides of nitrogen, in this equilibrium.

Supersonic aircraft fly in the stratosphere, much higher than subsonic airliners, because it is at this altitude, where the air density is some ten times lower than at sea level, that aerodynamic performance at Mach 2 and engine performance are at their best. And it turns out that the exhaust gases from the engines contain nitrogen oxides. So at the beginning of the 1970s, environmental activists were concerned about the possible harmful effects for the ozone layer that might be caused by a large commercial fleet of supersonic aircraft. Such a question could not be left unanswered and research began, significantly improving our understanding of the subtle stratospheric chemistry. But measurements had to be made directly in the stratosphere, and naturally, Concorde 001 offered an ideal platform for setting up the necessary heavy equipment. The latter, viewing through a lateral porthole of the aircraft, picked up the infrared light emitted by atmospheric nitrogen oxide molecules excited by solar light. Invaluable measurements were thus obtained regarding the quantities and reactivities of these molecules. Using these measurements and the amounts of gas ejected by the engines, chemists could then determine whether there was any significant risk. In this way, by the end of 1972, all the science goals, those of atmospheric chemistry and those of astrophysics, had been successfully brought together, whence the political decision-makers elected to maintain the prototype in flight through the first part of 1973. This was an unexpected and lucky convergence, which I gradually discovered during the friendly common meetings in which we set up the flight schedules.

One colleague, the austere but generous physicist André Girard, had long been pioneering the study of infrared light. The French government, extremely attentive to the possible harmful consequences of these notorious oxides for the Concorde program, had put him in charge of part of the study because he had designed a remarkably powerful and novel light analysis device, called a grid spectrometer. Viewing the Sun low on the horizon through the Concorde porthole, this device analyzes the Sun's light, part of which has been absorbed by the thick layer of the Earth's atmosphere it has had to cross to get there. It thus reveals the presence of nitrogen oxides. Science flights by 001 equipped with this instrument, in which André Girard took part to study the stratosphere, were thus carried out in the spring of 1973, alternating with tests of our own instruments in the hangar in Toulouse. Another colleague, André Marten from the former national telecommunications research center (*Centre national d'études des télécommunications* CNET), who would later join our laboratory in Meudon, had been specializing for some time in the study of planetary atmospheres, and particularly those of Jupiter and Saturn. So why not apply his research techniques to this problem of the terrestrial nitrogen oxides? He thus came to test his instruments aboard the Caravelle 116 at the flight test center, before setting them up in Concorde 001 in this same spring of 1973.

On 17 May, we were delighted to hear that we would be able to test the behavior of our instruments during a flight devoted to the study of the stratosphere that would take us from Toulouse right out over the Atlantic. This would be our first flight aboard 001. From my viewpoint during take-off, I could see the long pointed nose of the aircraft, starting in the lowered position and then lifted up in flight, while a visor system moved into place to obstruct part of the view from the cockpit when the plane accelerated. The large orange circle on the radar screen showed us the Bay of Biscay, which we soon left behind to break the sound barrier over the ocean, not wanting the supersonic "boom" to disturb the duck farms in the south west of France. Moving faster than sound in this westerly direction "brought the Sun back up above the horizon behind which it had just fallen," as our chief pilot André Turcat so nicely put it.[9] For the first time in my life, I was going faster than sound, and yet there was nothing to show that this was the case. It was a strange sensation,

---

[9] Turcat (2020).

one that thousands of Concorde passengers would later experience on their way to New York.

Even though there was no eclipse, the navigation had to be accurate. The plane's direction had to be held at 90° to the direction of the Sun, which was close to setting. The thickness of the atmosphere crossed by the Sun's light absorbs certain colors, in particular blue, as we see every evening when the Sun and sky turn an orange red, the color complementary to blue. Stratospheric nitrogen oxide molecules also absorb this solar light at certain precisely defined wavelengths, and it was the intensity of these wavelengths that the onboard measuring devices were designed to measure (Fig. 4.6). Once the measurements were made, we returned to Toulouse, while the great veil of night climbed to the east, the Earth's gigantic shadow on the pure atmosphere, a shadow I had so often admired from the observatories in Arizona and Chile, before my dialog with the stars. But here, in the perfectly clear terrestrial shadow, in a sky of unequaled purity, the horizon was curved, for at that altitude, one can witness the Earth's roundness directly.

The pilots, who showed a friendly respect for the rest of us, were very pleased with the first flight made in our company, whose navigation accuracy augured well for things to come. With his usual faultless vigilance, André Turcat was surprised by the way our delicate equipment stood up to the vibrations during the flight, and he congratulated us for that. Charles, who designed this feature, smiled modestly. We had just had our maiden supersonic flight, before a second rehearsal in May. We would have to dismantle everything to free up space in the cabin for a series of other tests, then put it all back in a few days just a month later.

The results of these studies of the stratosphere were eagerly awaited, but it was not until 1976 that the thick report by the stratospheric flight committee set up by the French government was finally published. And it turned out that stratospheric flights would have an effect on the protective ozone layer. But the measurements and calculations showed that the impact of a hundred Concordes in regular service would affect this layer by barely 1%, while the natural annual variability was already 20%, because ozone chemistry is also sensitive to other factors, such as the alternation of day and night or the 11-year cycle in the Sun's magnetic activity. So there was nothing to worry about! But science sometimes makes unexpected detours. Just as German occupation of Belgium had led Nicolet to investigate stratospheric chemistry, fears about the impact of supersonic flights made some look more closely at the behavior of this ozone layer. And in this same year of 1976, the publication of a report by the National Academy of Science in the United States caused new alarm. From earlier

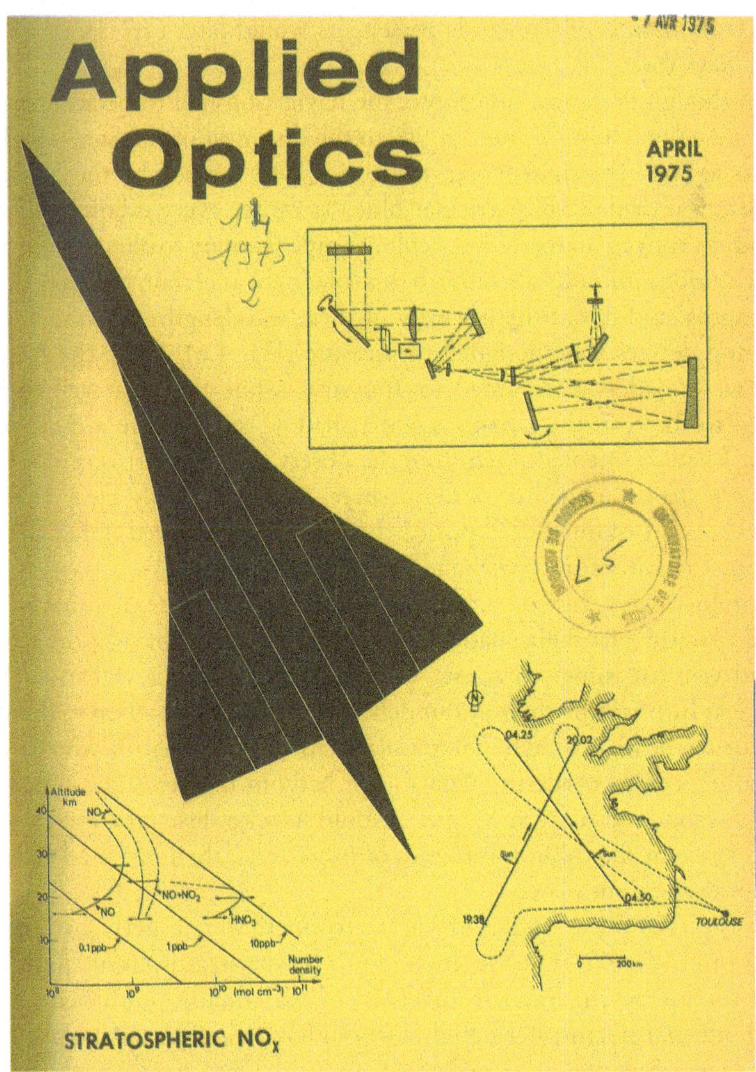

Fig. 4.6 The results of the observation of nitrogen oxides in the Earth's stratosphere, carried out by André Girard on Concorde 001, are featured in this issue of the scientific journal *Applied Optics*, which appeared in 1975. The consequences to be drawn regarding possible impacts on the ozone layer of a fleet of supersonic aircraft flying in the stratosphere would be the subject of an official French report in 1976. (Source: Applied Optics)

measurements made over a period of more than a decade, the report showed a constant and worrying decrease in the ozone layer that would be confirmed 3 years later by the formation of an "ozone hole" above the Antarctic continent. The reason for this decrease had been suggested back in 1974 by the American chemist Frank Sherwood Rowland, who had shown that chlorine- and fluorine-containing molecules used in vehicle air conditioners and private and industrial refrigeration systems, then released in large amounts into the atmosphere as a gas, were gradually destroying the ozone molecules.[10] Two years after the publication of the report, the USA and several other countries prohibited the production of these so-called chlorofluorocarbons (CFCs). Then in 1987, in an exemplary move, unfortunately rare at this level, the international community of 43 represented countries signed the Montreal Protocol which banned the use of such molecules. Sherwood Rowland (USA), Paul Crutzen (Netherlands), and Mario Molina (Mexico) shared the 1995 Nobel Prize in Chemistry for their work on atmospheric ozone chemistry.

It would be a good thing if, faced with the increasing concentration of carbon dioxide in the Earth's atmosphere, the major risks associated with climate warming, announced as early as 1976 in a report by the United States National Academy of Sciences,[11] would lead today to such radical and courageous measures on the part of the international community, for it challenges many private interests. But obviously, reducing the CO_2 in the Earth's atmosphere is more complex task than eliminating fluorocarbon gases.

4.8   Ready for Service

Four months had gone by since the green light of February 1973. Following confirmation of the flight plan, André Turcat obtained all the overflight permits required, particularly for the day of the eclipse, where we would have to fly over the Sahara, then Spanish (but administered by Morocco in 2014), Mauritania, Mali, Nigeria, and Chad, before landing in the latter. Only Mauritania refused at first. Indeed, this country was to host astronomy expeditions from the world over and had decided that, on 30 June, this part of the African sky would be closed to all air traffic. What a disaster it would have been if a plane had unexpectedly come between the eclipsed Sun and a telescope, or worse still, if a beautiful condensation trail (contrail) were to float by during a measurement made during totality! That is, unless some scientific sounding rocket should happen to collide with Concorde! But having convinced the authorities that Concorde's route had been determined with all the necessary precautions with regard to ground-based observation sites, thus

---

[10] http://en.wikipedia.org/wiki/Ozone_depletion.

[11] Charney (1979). https://nap.nationalacademies.org/catalog/12181/carbon-dioxide.

Fig. 4.7 This first day cover, of a kind so dear to philatelists, shows different scientific observations, on the ground, aboard Concorde 001, and using a sounding rocket, all carried out in Mauritania during the total eclipse

reassuring our colleagues who would be using those sites, Mauritania finally gave us the permit (Fig. 4.7).

Another of my astronomer colleagues, Pierre Charvin, held a position of responsibility in the Institut national d'astronomie et de géophysique (CNRS). From the very beginning of the project and helped by his colleague Michel Ravaut, he had been in charge of coordinating all aspects of organization and flight paths, thus relieving me with great skill from a task that I could not have done properly at the same time as preparing my own experiment aboard the plane. Aérospatiale had also allocated an enthusiastic and cautious engineer, René Joatton, from the Burgundy region like myself, to follow the scientific projects aboard Concorde and serve as the counterpart to Pierre for the eclipse mission. Diplomacy, insurance, and the press were thus admirably handled from start to finish. As head of mission, Pierre flew with us on the big day, as did his assistant engineer Michel Ravaut. Years later, he would become the president of the Paris Observatory and we would sometimes reminisce about our shared supersonic experience. June 1973 came, we had a rehearsal flight from Toulouse (Fig. 4.8), then flew to the Canary Islands.

Fig. 4.8 During the landing in Toulouse after the rehearsal flight on June 23, André Turcat, whose face still reflects the tension of these preparations, speaks with Pierre Léna in front of the press. (Source: Institut national de l'audiovisuel)

On 23 June, before leaving Toulouse for Africa, we held the first real rehearsal for all the astronomers and their instruments. A path was calculated that would take us far down to the south. Everything worked, except for our paper registering device, in which the pens became blocked, for in this distant period, such recording devices still used ink pens. Things could have been worse. We were ready for the real thing and took off for the Canary Islands on 27 June. A Caravelle came with us for technical support. It carried the large "Thermos flasks" containing our crucial and precious cryogenic liquids: dozens of liters of liquid nitrogen at minus 196 °C and liquid helium at minus 269 °C. It also carried a replacement engine for the 001, and all the technical teams for whom there was not room enough aboard the supersonic aircraft, and who would not therefore experience the breaking of the sound barrier (Fig. 4.9).

Fig. 4.9 Two days before the eclipse, the Las Palmas press marveled at the whole affair with the somewhat surprising headline: "In the shadow of the Sun's skin." Perhaps they were referring to the chromosphere?

Concorde landed at Las Palmas and excited much curiosity. The Spanish Guardia Civil watched over it day and night.

5

No Need for Alarm, It's Only an Eclipse

In the famous cartoon *Prisoners of the Sun*, one of the adventures of Tintin, these words are the calm declaration of Professor Calculus, although about to be burnt at the stake, as the Andean landscape suddenly went dark. On the morning of 30 June, as the Sun rose, already dented by the Moon, we were less composed than the celebrated savant with his pendulum, for any mishaps, no matter how small, might threaten the success of our great adventure.

5.1 General Rehearsal

On June 25 from Toulouse, we had a first rehearsal flight (Fig. 5.1). Then, two days before the event, on 28 June, we did a final rehearsal flight, taking off from Las Palmas and this time following precisely the flight path scheduled for 2 days hence, and even respecting the rendezvous times. Aboard the plane, each instrument was managed by a single astronomer. Stationed by his port-hole, he had therefore to deal with every issue that came up. Given the complexity of our own instrument, I was allowed three people to manage it: Alain Soufflot, Don Hall, and myself, proudly wearing our bright red flight suits. I did have to sort out a slight hiccup: for some reason, the technician who fitted our recording device, a stubborn chap from Britanny who came in the Caravelle, had imagined that he was going to fly in the Concorde. I had to set him straight and his reaction was somewhat churlish. I had to be extremely diplomatic in order to get the final adjustments of the device. These were crucial to us and he was the only one who knew them! Before taking off, standing below the great white bird under the tropical Sun, Charles

© The Author(s), under exclusive license to Springer Nature Switzerland AG 2025

P. Léna, S. Koutchmy, *Eclipsed Suns, the Solar Corona and Exoplanets*, Astronomers' Universe,

https://doi.org/10.1007/978-3-031-92199-5_5

Fig. 5.1 After the rehearsal flight of 23 June, standing by Concorde 001 on the tarmac at the airport in Blagnac near Toulouse, the pilots André Turcat and Jean Dabos (wearing flight suits), together with the science team. John Beckman proudly carries the delicate "Thermos flask" which had contained liquid helium, cooling his infrared-sensitive "thermometer" to a temperature of around minus 269 °C. (Source: F. Claudel)

Darpentigny and Jacqueline Mondellini helped us to pour the cryogenic fluid into the small and delicate cryostats containing the "thermometers," hypersensitive to infrared radiation. John Beckman did likewise. We place these on the frame that awaits them on board, let's start the pumps, check the alignments, and apply the brakes which must keep all this mechanics in position despite the vibrations of take-off. The acceleration is surprising in its suddenness, strong enough to hold us to our seats while the four jet engines roared and the afterburn tore us away from the ground.

Today the crew of six was precisely the one that would fly 2 days later. The pilots, André Turcat and Jean Dabos, were at the controls, while Michel Rétif, the test flight mechanic, was seated immediately behind them, keeping a close eye on everything, and especially the engines. Further behind was the test facility, a long row of cabinets packed with electronics and specific to the 001 prototype. There was a console on the cabin axis. Opposite these rows of dials and indicator lights were seated Henri Perrier, the test flight navigating engineer, Jean Conche who monitored the jets and their oil consumption, and

Fig. 5.2 The atmosphere aboard Concorde 001 during the eclipse flight, just before totality. In the foreground, Henri Perrier, the test flight engineer, is inspecting the map of Africa. Behind him is Michel Rétif, the mechanic, monitoring and checking all the flight systems. Further back, in the cockpit, darkened by the lifted visor on the aircraft nose, also lifted, are the pilots André Turcat (left) and Jean Dabos (right). (Source: P.L.)

Hubert Guyonnet controlling the navigation, all of them dressed in red. Through our headphones, we heard the exchanges between them on the onboard radio system (Fig. 5.2). We were accompanied by Jean-Pierre Aubertin, photographer and filmmaker. He was the only one whose presence was accepted by André Turcat during our flights, despite the many requests that were presumably made. Smiling and as discreet as a mouse, he produced the magnificent images that would eventually be made into the only film of the event, entitled *Eclipse 73*.

The stratosphere was still poorly understood, especially above Africa, where meteorological sounding balloons were rare and where observation satellites were still not available. Over the path that we would follow in pursuit of the Moon's shadow, the rehearsal flight of 28 June provided accurate measurements of the air temperature as a function of altitude. This was invaluable information for refining the schedule that the crew would specify to obtain a precise rendezvous with the shadow. Let again listen to André Turcat[1]:

> Take-off into the trade winds, retraction procedures for the landing gear, extinction of engine afterburns, half-turn toward the south, subsonic then supersonic climb, stabilization at a speed just above Mach 2, swing to the east to connect with the great circle (orthodromy) at the stipulated point on the map and within the expected quarter of a minute, all in half an hour cut into sections numbered in seconds. For the flight 2 days later, we had already decided to take off 20 s earlier in order to make up for atmospheric irregularities, and during this rehearsal on 28 June, I devised a procedure that would allow us to accurately drop a few seconds by applying the small airbrakes that can be used at Mach 2. These had been fitted on the prototype at my request right from the start. They had to be deployed for 10 s in order to lose 1 s, and this without interfering in any way with the engines, so as not to complicate the maneuver. We got back without mishap, nothing to modify aboard: we just had to await the big day.

When we touched down after 2 h and 36 min in the air, we were relieved for our instrument had worked well. We would just have to abandon one of the three infrared measurement channels whose "thermometer" had been contaminated when the Thermos flask (cryostat) containing it had been opened. Nothing dramatic. Our colleagues also seemed reassured, but we were concentrating so much on what would soon happen that we barely exchanged impressions during dinner at the hotel that evening.

5.2   Time to Go

The weather was stunning in Las Palmas on that morning of 30 June. The mild trade winds were blowing, causing the palm trees to wave, when suddenly a verse by the poet Paul Verlaine came into my mind[2]:

> Le ciel est, par-dessus le toit,
>   Si bleu, si calme!
>   Un arbre, par-dessus le toit,
>   Berce sa palme.

---

[1] Turcat (2020).

[2] *The sky above the rooftop/Is so blue and calm!/A tree above the rooftop/Waves its palm.*

This was the sky in which we were about to fly, the sky it had become my job to study, raising as it does so many questions about what lies beyond. The rendezvous was somewhere above the town of Atar in Mauritania. The navigator, Hubert Guyonnet, had specified it as 2033.1 N, 0900.5 W, 10 h 46 min 00 s UT, in the inertial guidance systems that would be used to steer the plane. At each instant, their gyroscopes would be the only way of knowing and informing the navigator where the aircraft was located above the Earth's surface. Our filmmaker came aboard. The film sequences he would produce were so unique and beautiful that this narrative alone could not replace them. I can only advise my readers to view the film *Eclipse 73* as soon as possible, and live our flight sequence for themselves.[3]

Up in the sky, the eclipse had already begun, and a dark indentation was encroaching slowly on the solar disk. For an immobilized ground-based observer, the Moon moves from one side to the other across the Sun's surface in 4 h and 30 min. Everyone was taking place in the plane and carrying out their final verifications. With all the necessary precautions, we had begun operations half an hour ahead of schedule. One of the inertial guidance systems was not working and had to be changed. There was too much kerosene on board and some of it had to be burned off by running the engines to lighten the aircraft. Just a minute before the take-off signal, a small aircraft was given the go-ahead to fly! It was 10 h08 (on the Universal Time scale, used throughout the following): the plane took off exactly on time, after an about-turn and the time to move down the runway, respecting the flight schedule established 2 days previously to within a second. It augured well. Our rendezvous awaited us in 45 min and 14 s. The shadow hastened along beside us. Coming from the west across the Atlantic at a speed of more than 2200 km an hour, it would soon reach the African coast. I fixed an air navigation chart to the side wall of the cabin. It showed the rendezvous, the time we should exit the shadow, and the predicted time of landing at Fort Lamy. I glanced at this occasionally to see how we were getting on.

In the words of our pilot once again[4]:

The atmosphere was less reliable than the previous day and made us lose 8 s over the climb. We just had to apply the planned maneuver to lose the remaining 12 s by which we were ahead of schedule. Three minutes before the rendezvous, we were flying 4 s behind the ideal time, well within the already ambitious toler-

---

[3] CERIMES (1973). http://www.youtube.com/watch?v=zHLyypLk-w and http://www.cerimes.fr/art-ciles/article_514/moment_771/plasma. See also the beautiful animated film of the 30 June flight at http://xjubier.free.fr/en/site_pages/solar_eclipses/TSE_19730630_Concorde001.html.
[4] Turcat (2020).

ance level, fixed at 15 s by the astronomers. But a challenge is a challenge, and since it was not a commercial flight, I decided to increase the Mach number by a few points beyond the so-called Mach maximum operating (MMO), taking us close to the risk of 'pumping' in the engine air inlets. It sufficed to monitor the speed without a moment's inattention, but we had already been doing that during the test flights. So in the end we came within 1 s and one nautical mile of our celestial rendezvous, not far from Atar, flying due east at Mach 2.04, with all the perfect stability of Concorde's characteristic cruise flight.

One or 2 min earlier, in the original viewfinder we had built ourselves, comprising a bundle of optical fibers and an eyepiece, I was watching carefully, now that the glare had become bearable to my eye, the image of the almost eclipsed Sun, located vertically above the plane, whose rays passed through the large porthole in the shape of a four-leaved clover, to be swallowed up in our instrument. The image-stabilizing mirror, controlled by a gyroscope and closely watched by Alain, worked extremely well as long as the plane was stable, with neither roll nor pitch. Suddenly, a red border appeared around the Moon's disk. This was radiation from hydrogen in the chromosphere. It was magnificent. Slowly, 15 times more slowly than for those observers now standing 16,170 m below, the lunar limb swept across the solar chromosphere and gave John Beckman the exact cross-section he had been hoping for. Not far from me, toward the back of the plane, Jean Bégot was guiding his telescope by hand and began to take a long sequence of photos. The instrument was hanging below the porthole in the cabin ceiling, and seeing him there clasping the long aluminum tube with both hands, he looked just like the lookout in a submarine, adjusting a periscope. I was dying to have a glimpse through the porthole, but there was no question of that, time was too precious. Concerning the photos of the spectacular events outside and the crew, totally focused on the task at hand, Jean-Pierre Aubertin was in charge of that.

The chromosphere disappeared. We reached second contact at exactly 10 h 53 min 14 s Universal Time. André Turcat had met the challenge quite magnificently, and Pierre Charvin made the announcement over the radio channel. In the plane and better than any ground-based observer, only two astronomers, Jean Bégot (Fig. 5.3) and Donald Liebenberg, could really enjoy the splendor of the corona through their instrument, when suddenly it came into view at the zenith, surrounded by stars, in a sky that was almost black at this high altitude. The crew and the other astronomers were unable to admire this stunning spectacle. For now began the infrared observation of this solar halo that had been seen by Plutarch and Thomas Edison. Fortunately, no

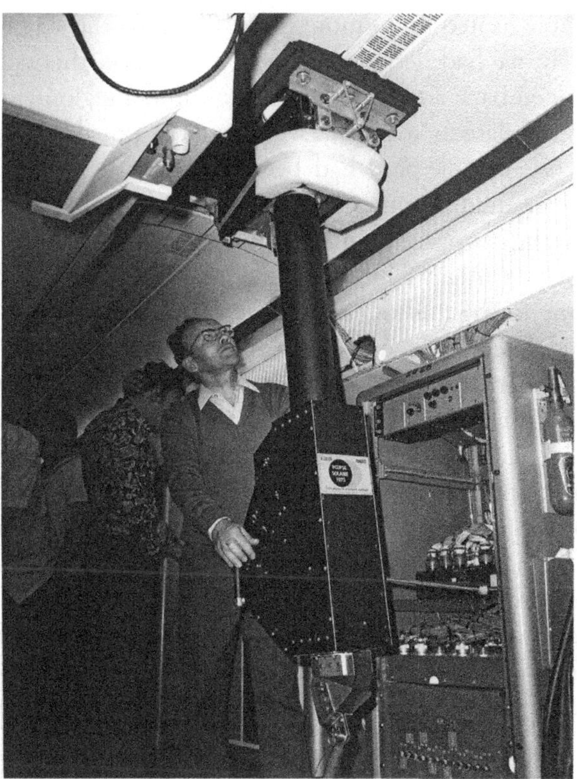

Fig. 5.3 Jean Bégot, from the team of Serge Koutchmy at the Institut d'astrophysique in Paris, manually orients his instrument, aiming at the eclipsed sun through one of the optical portholes. The instrument is mounted on a Cardan, allowing orientation, and elastic joints damp the aircraft vibrations. At the focus, three cameras where the incoming beam can be injected thanks to a manual control. (Source: S. Koutchmy, photography taken by J.P. Aubertin)

other flying object would be able to disturb us. A few moments before the second contact, I had already checked that the sighting was adjusted to a short distance from the solar limb, in fact, 1.5° of arc, with a carefully calibrated ruler. I thus switched on the motor that would slowly scan across to produce an image of the ring of dust, whose presence or absence our two "thermometers" was about to confirm. Over the radio, the recording of which would be crucial when it came to analyzing the measurements, I spelt out for Don Hall: "Pip alignment four [solar] radii. Pip five radii. Pip six radii. Pip seven radii. Restart: repointing on three radii. Pip … Pip for seven. Pip for eight. Excellent." Then a somewhat incongruous "Sh…t!' After the second contact, there was no longer any parasitic light from the solar disk and our "thermometers" were lit solely by coronal light, and by the intense infrared radiation from our porthole heated up to more than 100 °C by the supersonic air friction. Fortunately, we had designed a device to subtract this rather annoying signal. We scanned back and forth across the coronal region close to the Sun in order to accumulate as many measurements as possible. A glance at the paper register (whose ink pens had been checked!) where the intensity of infrared light measured by our "thermometers" was being recorded, and at 10 h 58, I made the happy announcement: "We've seen the dust [of the rings], we've seen it." Under Alain Soufflot's vigilant eye, Don Hall and I would roll out our program during this totality that would exceed 1 h on the clock. Now sure that we had detected the much hoped for infrared signal from the ring of dust, we began to analyze it with a spectrograph to determine the composition of the microscopic grains making up the ring. Under the stars at high noon with all its navigation lights on, Concorde hurtled on due east into the night. We renewed the magnetic tape of the register. Pierre Charvin announced over the radio: "Third contact in 6 min." The last few scans, almost a routine soon to come to an end. And 3 min and 25 s before this third contact, we had ended our measurements and dashed to the side portholes of the cabin to marvel at the extraordinary spectacle and finally get our own photos (Fig. 5.4). I am almost ashamed to say here that we thus neglected our measurements, even though completed, for 3 min stolen from the totality, a length of time

Fig. 5.4 During totality on 30 June 1973, an extraordinary spectacle was observed through a side window of Concorde 001 at an altitude of 17,000 m. In the foreground, the "gothic" delta wing of the plane shows a faint reflection of the horizon. In the background, the glowing horizon reveals the curvature of the Earth. To the south, on the African landscape yellowed by drought, there is a clear boundary between the umbra and the penumbra. Far away, the lower atmosphere is milky white, scattering the Sun's light. Higher up, above an altitude of 12 km, the stratosphere scatters a beautiful blue light (Source: P.L.)

for which most astronomers are ready to mount expeditions to the other side of the world!

A part of Africa had been plunged into darkness by the huge shadow cast upon it by the Moon, within which we were still flying. Beyond the edge of the shadow, toward the south, the savanah, yellow with drought, stretched away into the penumbra, as far as the eye could see, where distant towers of cumulonimbus threatened storms in the intertropical convergence zone above Niger. In the foreground, the white gothic wing of the aircraft reflected a glowing horizon, rounded by the curvature of the Earth. Above the horizon, far away and hence lit up by the Sun, the Earth's atmosphere scattered light, milky white in the lower layers of the troposphere where sandstorms had whipped up dust, then darker and darker blue as one looked higher into the stratosphere, until one became lost in the darkness of night. Jean Bégot, too, was taking photos of this extraordinary scene, and one of them would soon come to the attention of the tabloids.

After the third contact at 12 h 07 min 13 s, the lunar disk once again revealed the reddish glow of the chromosphere to John Beckman. We were flying at 17,600 m because the plane, lighter now that it had consumed a good part of its kerosene, had been climbing ever higher into the stratosphere. Pierre Charvin passed me a paper that I still keep carefully after all this time. It confirmed that we had enjoyed 74 min of totality, almost exactly ten times the maximal duration of an eclipse when it is observed at ground level. We had just beaten the record, observing totality for longer than the cumulation of all eclipses viewed for a quarter of a century. As Pierre Charvin noted later on: "Those who had christened this the longest eclipse of the century didn't realize just how right they were".

Throughout the flight, Hubert Guyonnet, the radio navigator, answered questions raised by the pilots: "Our ground speed?—1102 knots." (This would be 565 m/s, or 2037 km an hour.) "Predicted time of arrival at 23°36′N?—10 h 27." "Distance to Niamey?—480 nautical miles." "Wind?—46 knots from 087 [an east wind]." Those were the 46 knots of opposing wind that had prevented us from reaching the hoped for 80 min of totality. From Jean Dabos, the copilot, André Turcat was asking mainly for weather reports at Niamey, Tamanrasset, and Kano, which were alternative landing sites in case of rerouting, until they were no longer attainable, in order to be able to make decisions soon enough on the basis of progress information provided by Henri Perrier and Hubert Guyonnet. But in the end the weather was fine everywhere, perhaps too fine for all these thirsty countries, and the threat of the intertropical weather front kept well away from us (Fig. 5.5).

Fig. 5.5 On this stamp, issued by the Republic of Congo in 2009 to commemorate the International Year of Astronomy, Concorde is still remembered, as the commercial airliner wearing French colors flies before the terrifying cumulus clouds that characterize the intertropical convergence zone

The light from the solar disk was returning little by little, and 12 min after third contact, our colleagues' final observations were completed and the plane went into a bend toward the south. Forty years later, our navigation mechanic Michel Rétif wrote to me in the following words[5]:

The engine regime was being reduced when suddenly one of them, instead of stabilising and slowing down, actually stopped altogether, without any alarm going off. In this situation, in supersonic flight, the non-functioning engine begins to rotate on its own due to the air flow, and this has the effect of maintaining the electricity supply to the aircraft. In subsonic flight, the consequences of such an incident would have been spectacular, with the corresponding loss of electricity generation. But this engine restarted very quickly and subsequently accepted the slowdown.

---

[5] Michel Rétif: Personal communication, 2013.

We locked our equipment in place and went back to the landing seats, ready to touch down on the runway at Fort Lamy in Chad some 49 min later. To check that the runway was indeed clear, and despite the fact that there was not much fuel left in the reservoirs, André Turcat made a first low altitude transit over the site. The weather was magnificent, the Sun still half eclipsed, and a kind of euphoria came over me as the wheels of the Concorde touched the ground in Chad. The exit steps were put in place and we climbed down, still wearing our red suits. In a flash, I thought of my father who, now aged 81 and always enthusiastic about exploration in Africa,[6] would be eagerly awaiting a telegram in Paris to announce our success.

The Caravelle carrying all our other colleagues was much slower and would only arrive at Fort Lamy during the afternoon, long after us.

5.3   A Visit to the Nyangatom People of Ethiopia

Traveling at more than 2000 km an hour, the Moon's shadow whose pursuit we had just abandoned continued on its way. It crosses Kenya where several teams from the USA are based, including Gordon Newkirk, then it brushes past Ethiopia and the southern tip of Somalia, before disappearing across the Indian Ocean (Fig. 5.6).

The Nyangatom are an extremely isolated people living in the region between southern Sudan and Ethiopia. Two French ethnologists, Serge and Marie-Martine Tornay, paid them a visit. They warned them about the eclipse that was about to occur. Every month a soothsayer anoints the villagers with white clay, sprinkling them with saliva and chewed up aloe leaves, in order to bring back the Moon at the time of the new Moon. Let us turn to the ethnologist Gérard Francillon,[7] but using Universal Time rather than local time for easier comparison with the story we have told so far:

---

[6] The demonstration for using gazogen powered trucks, carried out by Henri Léna in 1929, following approximately the same route as Concorde 001, is presented in (Léna 2018).

[7] The quote comes from a research note, published by Gérard Francillon in 1974. The full analysis of the ethnological observations made in Ethiopia is given in (Tornay 1979).

Fig. 5.6 On 30 June 1973, the lunar shadow brushed past the southernmost point of the Federal Republic of Somalia as it left Africa. This unfortunate country, affected by a devastating drought and governed by a remorseless dictatorship, would gradually sink into tragedy

13 h11. The soothsayer continues the sprinkling. He spits upon all the women in the hamlet who turn up. The eclipse seems to be at maximum. A woman is heard to shout: 'Darkness, all is darkness!' Only then do the others notice that something unusual is happening. The soothsayer examines the Sun and confirms that it is partially dead.

13 h15. The light is gradually returning. Suddenly someone shouts: 'Strike the gourds, strike the gourds to bring back the Sun. Hit them hard, it's coming back, it's like the Moon.' The women then begin the necessary pounding and pummeling.

13 h21. The noise dies down. We return to the soothsayer. 'Did you see?' he asks. 'It has returned and it seems contented.'

14 h16. The country has recovered.

The cultural background of these inhabitants of Ethiopia provides them with a ritual response to the event. For them, the cosmic phenomenon is viewed as a sign, rather than as an event.

Fig. 5.7   In the lower Omo valley in Ethiopia, 5 min after totality during the eclipse of 30 June, this Nyangatom woman strikes an upside-down milking pot. Photographed by the ethnologist Serge Tornay, she turns her back to the Sun (Source: G. Francillon)

So concluded the ethnologist Gérard Francillon,[8] while the shadow went on its way (Fig. 5.7).

---

[8] Gérard Francillon in *Note de recherche: Eclipse de soleil du 30 juin 1973* (1974), unpublished. A more detailed analysis is in (Tornay 1979).

5.4   Mach 2.05 or Above?

Almost 40 years later, in a moving letter sent to me just before he passed away, our flight engineer Henri Perrier alluded to his indictment after the tragic accident in Gonesse, before the final judgment in which the charges against him were dropped. He also gave me his own version of our race against the Moon's shadow, which I record here[9]:

> Among my recollections of these rehearsal flights, and then the flight itself, two of them remain vivid: the accuracy in the specification of the time to within one tenth of a second [the click of the clock], while the best estimate of our geographical position using the inertial guidance systems was between 1 and 1.5 nautical miles (i.e., equivalent [in time] to between 3 and 5 s); and the fact that, in order to compensate for a wind component along the flight path that was greater than had been registered during the flight on 28 June, Michel Rétif and I, with the unspoken agreement of Turcat and no longer restricted by the external observations, had decided to cruise at Mach 2.05 by removing the [automatic] control systems which reduced the thrust as soon as we went above Mach 2.01–2.02. To do this, on the chart recorder at my post, I could observe the significant pressures of the surge margins and use these to tell Rétif what manual display [to choose] in the position of the air inlet ramps. This was a 'reasonable' operation as long as our trajectory was not expected to subject us to sudden changes in static temperature, according to our rather vague knowledge of the weather in this part of the atmosphere and at these latitudes. Having spoken of this with my old boss [Turcat], I know that he had no recollection of the risk taken here with regard to the jet pump, which would have been impossible to make good on the prototype without significant deceleration, and this to gain perhaps 1 min of observation.

I have kept the recordings of the flight parameters just as they were stored by the aircraft test facility. I would read the Mach number minute by minute during the flight. And there was no doubt: we went up to Mach 2.09 for several minutes! According to this same recording, our ground speed at this precise time was 551 m/s or 1983 km an hour (Fig. 5.8).

Did we actually fly faster than the shadow? That would be another record, although of a more sentimental nature. The shadow never moved more slowly than 601 m/s, and our average ground speed throughout totality was 568 m/s, but with variations that no doubt took us very close to 601 m/s. If only the wind had not been against us! The professional flight log of André Turcat does not cover such fine details (Fig. 5.9).

---

[9] Henri Perrier: Personal communication (2011).

TEMPS	ZP (M)	MACH	TA (DK)	TETA (DEG)	PHI (DEG)	VSOL (M/S)
11-35-10.00	17356.	2.10	180.1	64.4	0.2	-551.
11-35-11.00	17356.	2.09	180.5	64.4	0.2	-551.
11-35-12.00	17357.	2.09	180.9	64.4	-0.1	-551.
11-35-13.00	17357.	2.09	180.9	64.3	-0.1	-551.
11-35-14.00	17356.	2.09	180.3	64.4	-0.1	-551.
11-35-15.00	17356.	2.10	180.0	64.4	-0.1	-551.
11-35-16.00	17356.	2.10	179.9	64.3	-0.1	-551.
11-35-17.00	17356.	2.10	180.1	64.4	-0.1	-551.
11-35-18.00	17356.	2.10	179.9	64.3	-0.1	-551.
11-35-19.00	17356.	2.10	179.9	64.4	-0.1	-551.
11-35-20.00	17355.	2.10	179.7	64.4	-0.1	-551.
11-35-21.00	17355.	2.10	179.7	64.2	-0.1	-551.
11-35-22.00	17355.	2.10	179.7	64.4	-0.1	-551.
11-35-23.00	17355.	2.10	179.7	64.4	0.2	-551.
11-35-24.00	17356.	2.10	179.8	64.2	-0.1	-551.
11-35-25.00	17356.	2.10	179.9	64.2	0.2	-551.
11-35-26.00	17357.	2.09	180.6	64.4	0.0	-551.
11-35-27.00	17357.	2.09	180.7	64.3	-0.1	-551.
11-35-28.00	17357.	2.09	180.7	64.4	-0.1	-551.
11-35-29.00	17357.	2.09	180.9	64.4	0.2	-551.
11-35-30.00	17357.	2.09	180.9	64.4	0.2	-551.

Fig. 5.8   Extract from the onboard recording of 30 June, used to determine the exact trajectory of Concorde 001. It is 11 h 35 min 10 s and the plane is flying at 17356 m at a speed of Mach 2.09–2.10. As the Mach number is calculated from the local air temperature at this altitude, this number varies, even though the ground speed shown in the last column remains constant. (Source: P.L.)

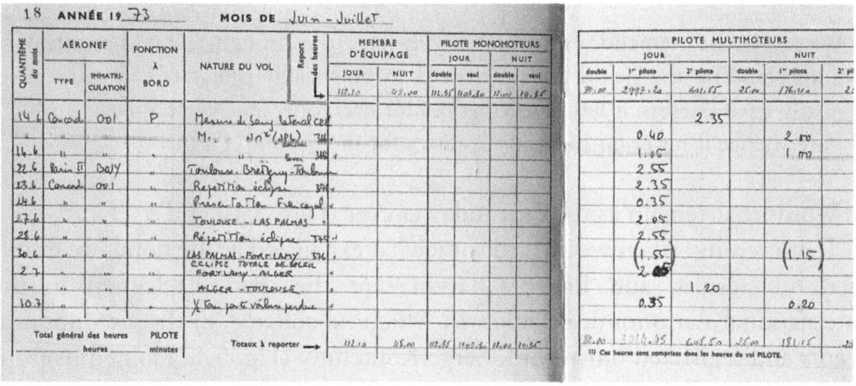

Fig. 5.9   A partial section of André Turcat's personal flight log (June and July 1973). The eclipse flight dated 30.6 is marked with 1.55 h of daylight and 1.15 h (i.e., 75 min) of night, corresponding to the totality. (Source: Document kindly communicated by Francette Joanne)

5.5   A Failed *Coup d'Etat*

Fort Lamy appeared to be filled with scenes of jubilation. We were coming down the main avenue with André Turcat in the lead, surrounded by crowds of people, many of them carrying blackened glasses to watch the eclipse, as totality had ended more than an hour before. With some surprise, we noted machine gun posts set up at certain crossroads. The explanation was soon made clear by Mr. Baldit, the representative from the French embassy who had come to greet us. There had just been an aborted *coup d'état*, but we did not know whether its architects, castigated by the article quoted below, had deliberately chosen to make it coincide with a solar eclipse.

Two days later, on 2 July, the Chad press agency published the following slightly pompous account in its roneoed bulletin which I still keep, under the title *Le Canard déchaîné et les conspirateurs*[10]:

> Chad has been living through dreadful times since achieving independence [...]. If the Chadian people have been able to stand up to this trial, it is thanks to the composure, and above all the unbending patriotism of François Tombalbaye [the president]. Just a little further and Dopélé would have plunged us last week into the midst of the most catastrophic events, and by now Chad would be a blazing inferno or a gigantic field of dead bodies.

So we had a narrow escape! This same issue of the newspaper tells of the arrival of Concorde and quotes André Turcat's remarks to the journalists:

> We know the tremendous problems you face [the drought], but the task of understanding the world does not need to be socially motivated in order to justify itself. Research has consequences for everyone, and what appears to be pure science is never far from applications for the benefit of all.

Words that remain as relevant today as ever (Figs. 5.10 and 5.11).

That evening, when we arrived somewhat exhausted at the hotel, we examined our results, and Pierre Charvin sent the crucial telegram to the International Astronomical Union in which he collated all the astronomical events and registered our record. Serge Koutchmy (Fig. 5.12) joined us from Moussoro, kindly transported by a Puma helicopter from the French army.

---

[10] Chad Press Agency, Daily Bulletin: *Info Tchad*, no. 2809, 2 July 1973 (roneoed). The title is a play on words, *Le Canard enchaîné* being a well-known satirical newspaper in France (the word "*canard*" is slang for a newspaper).

Fig. 5.10 André Turcat (left) after the landing at Fort Lamy. Facing the camera, one of the Aérospatiale engineers. Behind him, Michel Ravaut and at right Donald Hall wearing a red suit. (Source: P.L.)

Fig. 5.11 Reception in the town of Fort Lamy. In the foreground, left to right, the scientists Yves Viala, Jean Bégot, and Michel Ravaut, accompanied by a small crowd of onlookers. The eclipse had just ended. Shadows are short, the Sun being close to zenith (Source: P.L.)

We also contacted our research institutes, which would have to answer questions posed by journalists. At first glance, all the instruments seemed to have worked well, we had harvested a considerable amount of data on the corona, and we were fully satisfied with the way things had gone. So many

Fig. 5.12 Ambiance of a ground eclipse team, in Mexico (1970). Facing other solar astronomers, Serge Koutchmy is gathering data for his PhD. From left to right, Jay Pasachoff, Zadig Mouradian, Marius Laffineur, and Raymond Michard. (Source: S. K., image kindly provided by Mrs. Mouradian)

unexpected incidents might have brought failure to our mission: we might have run out of cryogenic fluids, poor weather might have forced a reroute, the recording devices might have broken down, a porthole might have burst, vibrations might have gone out of control, or there might have been a problem with an engine air inlet.

5.6   The Return Home

Two days later, the 001prototype left N'Djamena for Toulouse, stopping over for 6 h in Algiers, where we took members of the government aboard. The correspondent of *Le Monde* made the following observation: "The Algerian authorities were visibly sensitive to this mark of respect." Its duty accomplished, the Caravelle had made a stopover on the island of Djerba in Tunisia, and the whole team had bathed in the mild waters of the Mediterranean. After dismantling our observation instruments, not without some sadness, each of us left Toulouse, taking our measurement results with us for analysis. Apart from myself and my friend Don Hall. On the evening of our arrival, Don was struck by a terrible African fever and had to be hospitalized in Purpan. In his room, I watched over him with great concern, but the robust constitution of his 28 years, celebrated only the day before the eclipse, soon had him out of danger, although 10 kg lighter.

After its last flight on 19 October, the Concorde 001 prototype, under the matriculation F-WTSS, came to land at Le Bourget, having accomplished a grand tour over the Atlantic. On board was an 82-year-old lady, the Hungarian countess Lilly von Coudenhove-Kalergi,[11] who was one of the first women to obtain a pilot's license, back in 1912. She was so enthusiastic about this plane that Aérospatiale offered her the 397th and last flight made by the prototype. Concorde was handed over to General Lissarague who was the director of the brand new *Musée de l'air et de l'espace*. Formally and symbolically, André Turcat was to present the keys of the plane to the general. But unfortunately, there is no key for a test plane, so caught short, and not wishing to disappoint, Turcat took his car keys out of his pocket! Concorde 001 had flown in supersonic flight for a total of 812 h and 9 min. André Turcat, who was the first and the last to fly this extraordinary prototype, made the following address[12]:

It made too much noise and too much smoke. It didn't have sufficient range to cross the North Atlantic. It was jam packed with test equipment. It didn't have that watch with 24-h dial which doesn't tell anyone the time, but which is required by certain airlines. And to be quite honest, the ash trays were all full. The only thing left for this old chap, already 5 years old, is clearly the museum. It achieved quite a lot though. It was the first to carry 140 tons in supersonic flight. It flew from 200 to 2300 kph. It crossed the South Atlantic in 2 h and went beyond the 45th parallel south. It climbed, nose-dived, and rolled, and it was pushed and mistreated as much as any other plane. It underwent thousands of modifications. It was a docile subject in the hands of ten test pilots and 20 airline pilots. It was admired for its power and elegance, now surpassed by its successors. It carried the most famous of people. It spent more than an hour under the total eclipse of the Sun, a record that will take a very long time to beat. It measured the nitric acid in the stratosphere and watched the shadow of the Earth on the depth of the sky. It gave us the greatest joy of our careers, and it announced a new era in world transport. In a word, we simply adore it.

Concorde 001 is still at the museum in Le Bourget, joined by F-BTSD, one of the commercial airliners of the Air France series (Fig. 5.13). At the very front of its fuselage features the logo of the African eclipse and, on the ceiling of the cabin along which thousands of visitors now walk, the four portholes are still there.

---

[11] http://www.pionnair-ge.com/spip1/spip.php?article441.
[12] Turcat (2020).

Fig. 5.13 At the aeronautics museum in Le Bourget (Seine-Saint-Denis, France), 40 years on, the Concorde 001 prototype still carries the eclipse logo from 1973 on the side of the fuselage (Source: Musée de l'Air et de l'Espace)

6

Seventy-Four Minutes of Observation, But What Gain for Science?

It's all very well setting up a record observation time of a solar eclipse, but that wasn't really the main objective for the scientists who set off on this adventure. One might say that it was only a secondary result. It was the Sun and its corona that really interested us. So what discoveries were made thanks to the three exceptional circumstances we benefited from: the long period of totality, the slow scan of the chromosphere just before and just after totality, and the excellent access to infrared emissions from such a high flight path?

6.1 Where We Search for Dust Rings in the F Corona

So are there rings or shells of interplanetary dust concentrated around the Sun in the plane of the ecliptic, that is, in the plane containing, to a good approximation, the orbits of all the planets? Our predecessors made this assertion on the basis of rather difficult ground-based observations during the eclipse of 1967. After carefully analyzing our observations of the infrared light from the corona (Fig. 6.1), Don Hall and I reasserted this claim in a talk based on our results and presented in São Paulo at the end of 1973. However, as often happens in science, things turned out to be much more complex than we had assumed in those days of innocence, and the subject led to much debate.

The original version of the chapter has been revised. A correction to this chapter can be found at
https://doi.org/10.1007/978-3-031-92199-5_10

P. Léna, S. Koutchmy, *Eclipsed Suns, the Solar Corona and Exoplanets*, Astronomers' Universe, https://doi.org/10.1007/978-3-031-92199-5_6

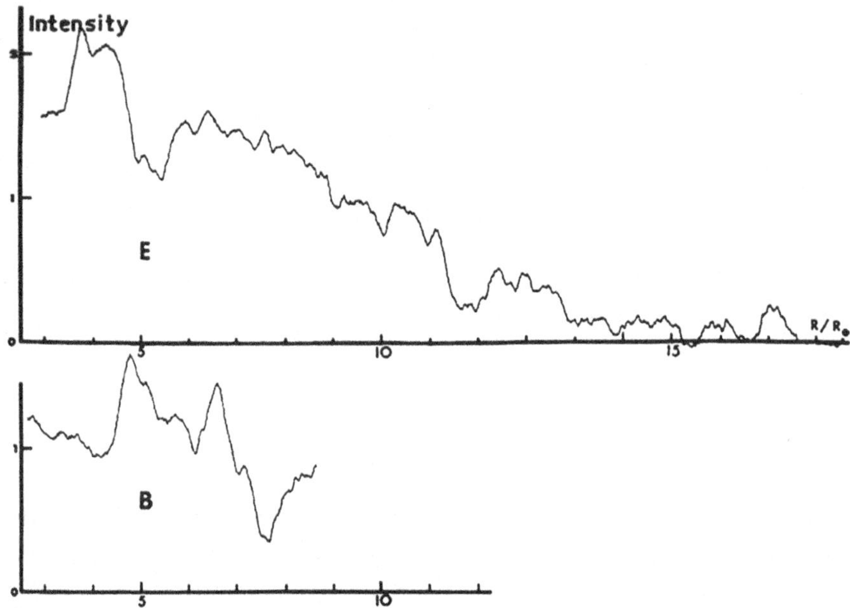

Fig. 6.1 Original recording of infrared emission by dust shells, as we considered we had detected them during totality in the flight of 30 June. Each of the two curves shows a measurement of the infrared radiation intensity from the F corona. These two measurements were made during totality, moving away from the center of the Sun along the ecliptic, through a distance of between about 4 and 15 apparent solar radii (the apparent solar radius seen from Earth is 0.25°). (Source; P.L./Observatoire de Paris & Astronomy and Astrophysics)

Fifty years have passed, as I write again about this true story of our eclipse, and following a great many observations made from high mountain tops, such as the Hawaii observatory at 4200 m altitude,[1] or from a Japanese stratospheric sounding balloon, the results remain contradictory across the vast numbers of published papers, despite the fact that each piece of research is of the highest quality. Some also measured infrared emission from the corona, at a distance of a few solar radii from the center of the star, while others, just as respectable and conscientious, observed nothing of the kind, although making their images with modern infrared cameras. This already shows that, perhaps with a certain naivety, we didn't choose an easy problem when we selected such a complex measurement program for our Concorde flight. From these apparently contradictory results, it now seems possible to draw two quite different conclusions: either all those who, like ourselves, measured coronal infrared emission suggesting the existence of a dust ring were completely

---

[1] Lamy and Gilardy (2022).

mistaken, for example, because their instruments were not sensitive enough, or suffered from unidentified background signals, or they misinterpreted those measurements; or else the rings do exist but not all the time, something which might reconcile the two series of observations.

Clearly, the second conclusion appeals much more to me than the first, even though, as a scientist, one must always be ready to admit error. From the 1980s, and for many reasons, my work drifted away from the corona and its dusts and I ceased to work on the problem. But in 2010, Lyubov Shestakova, an astronomer at the Fesenkov Astrophysical Institute in Kazakhstan, published an article that attempted to reconcile the two possibilities, suggesting the possibly sporadic existence of circumsolar rings.[2] But other specialists, publishing practically at the same time, consider all doubt here to have been removed.

We will return to this later because, 50 years after our measurements, the F corona or dust corona, is far from being known yet, although it is much better known than in 1973. Its relevance therefore deserves what we present to the reader in Part II of this book.

For their part, physicists calculate the behavior of a tiny rocky grain of interplanetary material—measuring just a fraction of a thousandth of a millimeter—when it is subjected to the Sun's gravity, which attracts it, while receiving light which tends to push it away from the Sun, heating it, and in some cases causing it to vaporize. Furthermore, various kinds of crystal grains, well known by dust gathered from shooting stars in our atmosphere or even in space, will display quite different behavior depending on whether they are made from iron-bearing or magnesium bearing olivine, for example, to cite only this particular mineral, a silicate. So the question here is not whether such grains transit in the immediate neighborhood of the Sun, before they are heated up to more than 1500 °C and evaporate, but whether these grains can concentrate sufficiently to form rings or shells.

6.2 How to Heat the K Corona to a Million Degrees

At the time, a more fundamental question than the nature of the rings bothered solar specialists. The surface of the Sun, which is what we see in the sky and which sends us its beautiful white light, is not really at a very high temperature, in fact, barely 5000 °C, nothing compared to the millions of degrees in the Sun's core. Above this surface, the temperature decreases slightly—it

---

[2] Shestakova et al., https://arxiv.org/abs/1003.2818.

was this decrease that I was studying aboard the NASA Galileo I aircraft—but then it increases suddenly, and the hydrogen and the helium in this region a few thousand kilometers higher are already at 20,000 °C. Consequently, hydrogen atoms are broken in electrons and protons, and helium atoms loose one of their two electrons. It is the light from this thin layer, called chromosphere, and briefly swept by the lunar limb when viewed from the ground, which produces the so-called red flash, which ancient myths interpreted as the blood of the Sun when it was eaten by the Moon, just before and just after totality. Barely 1000 km higher, the gases begin to rarify and this is the corona proper, where the temperature goes on increasing until it very quickly reaches a million degrees. This sudden increase in temperature in the chromosphere and the corona is not so easy to explain: light coming from the Sun surface, containing neither significant ultraviolet nor X-rays, cannot heat this gas because it is much too rarified to absorb it, so one would expect the temperature to fall off with altitude. But what then could be the cause of this heating? This was, dare we say it, the burning question which preoccupied Donald Hall as much as Serge Koutchmy or John Beckman when it suddenly became possible to carry out a lengthy observation of this region from Concorde.

If light could not heat the gas, perhaps some other waves could do so, dissipating there the energy they carry. But what kind of waves? When we speak of waves, we think of oscillations, time varying quantities, like the vibration of a violin string or a wave on the surface of the sea. Now it was precisely the slowing down of the Moon's motion relative to the Sun due to our supersonic flight that gave time for time to pass either at second contact when the Moon begins to cover the chromosphere or at third contact when it reveals it again, or during totality, when there would be plenty of time to observe any periodic light variations that such oscillations would be sure to produce in the corona. In fact, it has been known since the 1950s that sound waves, also known as acoustic waves, are permanently present in the lower layers of the Sun, and their period of oscillation has been established as 5 min. This was a major discovery which, even half a century later, still inspires study of the Sun, and today also the stars. After his flight in Indochina, Raymond Michard, mentioned earlier as head of the Paris Observatory in 1972, was one of those who made this discovery during a visit to the USA. Could it be that these acoustic waves, born in the depths of the Sun, propagate toward the corona and transform into violently non-linear shock waves, which then locally dissipate the energy they transport into heat, thus heating the coronal gas? An attractive hypothesis that the Frenchman Evry Schatzman, among other famous astrophysicists, put forward in 1949 and which deserved to be tested at supersonic speed! Faced with these hypotheses, it is therefore not surprising that two of the major observations scheduled aboard Concorde, that of Serge Koutchmy with Jean Bégot and that of Don Liebenberg from Los Alamos, were intended to provide elements of answer.

The Program of Serge Koutchmy and Jean Bégot

Searching for the trace of acoustic waves was the central question of Serge Koutchmy's program, which Jean Bégot carried out under his porthole. He obtained 250 color photographs of the corona (Fig. 6.2), as well as 36 black-and-white of better sharpness, since he could switch between the two modes during the 74 min of totality.

By completing their analysis[3] with that of photos obtained on the ground during the same eclipse by Serge Koutchmy in Moussoro (Chad), the hypothesis of acoustic waves could not withstand for long the examination of the whole sequence of successive images and had to be abandoned during the following decades. However, the images, at the limit of the sensitivity of the photographic films used, revealed structures in the polar regions of the

---

[3] Koutchmy (1976).

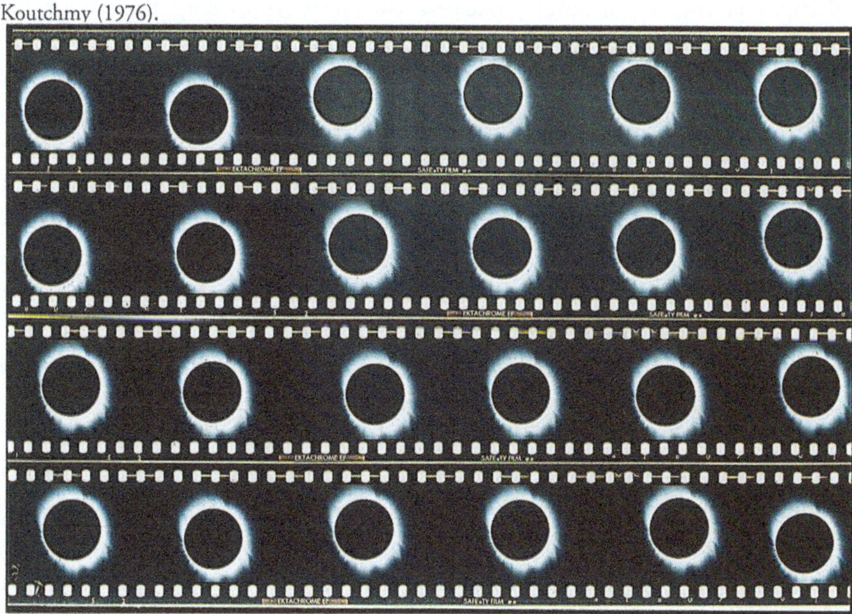

Fig. 6.2 A sequence of color images during totality, obtained by Jean Bégot, every 20 s, exposure time 1/60e of a second. The hand pointing being not perfect, the position of the image slightly varies. At this scale, the perturbation of the image quality by the supersonic airflow cannot be perceived. (Source: S. Koutchmy)

Sun—called plumes or polar jets—of periodic oscillations, with periods ranging from 100 to 1000 s. Later, working at the famous Sacramento Peak solar observatory (New Mexico, USA) equipped with a coronagraph, Serge Koutchmy demonstrated, on a spectral line of iron, faster oscillations with a period of 43 s.[4] The observation made during the Concorde flight therefore anticipated the discovery of another type of rather transverse wave, called magnetohydrodynamic waves (MHD), the existence of which was firmly established later and to which we return in Chap. 7. Unfortunately, these images of the corona, taken from inside a supersonic plane, when the light had to pass through an extremely disturbed layer of air, did not have the ultimate quality hoped for and their authors were not able to establish as solidly as necessary which was a magnificent discovery. This blurring of the images, here taken in flight but which astronomers encounter in many other circumstances, will merit a brief analysis below.

Blurred Images

Perceiving the finest details of an astronomical image is one of the difficult challenges facing astronomers, because the light producing the image has encountered all kinds of avatars on its path: it has passed through the atmosphere more or less agitated by the Earth, it has been reflected by mirrors and passed through lenses which are never perfect. The first cause is called *seeing* by astronomers, it is also at the origin of the well-known scintillation of stars.[5] Observing from Concorde adds the fact that the instrument, attached to the plane which pitches and rolls slightly, does not always point exactly in the desired direction. In addition, a layer of air is crossed, called the boundary layer, where air velocity increases from zero to more than Mach 2 over a few tens of centimeters in the vicinity of the skin of the aircraft, itself heated to more than 100 °C by the friction of this air. Examination of an instantaneous and highly enlarged photograph, obtained by Jean Bégot during the supersonic flight, clearly shows variations in the sharpness of the lunar edge depending on the portion examined (Fig. 6.3). This variation reflects this disturbance introduced into the image by the crossing of this layer. Measuring this effect to better analyze the images is possible, but does not restore the quality they would have in the absence of a window. Far from astronomers, engineers who

---

[4] Koutchmy et al. (1983).

[5] The curious reader would find the story of the seeing and its modern countermeasures on giant telescopes in Léna (2020).

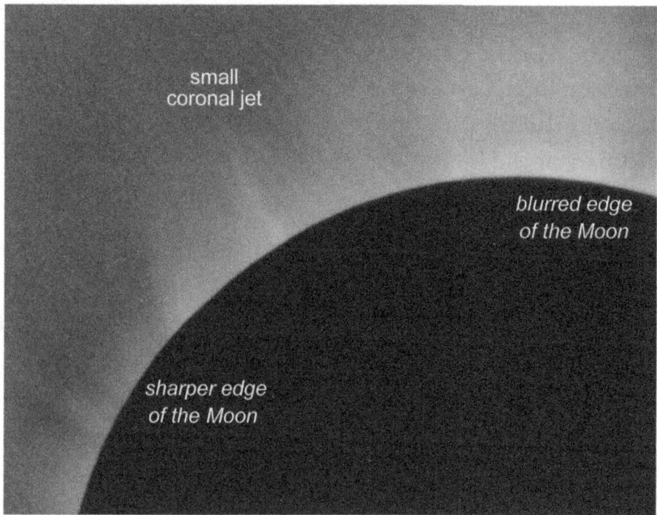

Fig. 6.3   Fine details in a blown-up image of the corona, taken by Jean Bégot during the totality. The sharpness of the Moon's edge varies, while coronal features are conspicuous. (Source: S. Koutchmy)

design supersonic missiles and their laser guidance devices must take into account this disturbance of the optical quality, caused by the boundary layer.

The Los Alamos Program

We have already met Donald Liebenberg, this veteran of eclipse observations by airplane, modest, and smiling. He built a beautiful instrument for Concorde, using a television camera to film the corona. Filtering the light with a device a filter Perot-Fabry, he very precisely isolates in the light of the corona a wavelength in the green, emitted by iron atoms having lost 13 electrons, therefore having become ions. Only a gas reaching a temperature of two million degrees Kelvin can emit this spectral line: it therefore provides an exact signal of the hot gas presence, without other cooler regions being able to disturb the measurement. This precise filter only accepts a narrow field of view, so it is limited to a portion of the corona, chosen near the solar equator and close to the disk, where the corona is the brightest. Despite some spurious signals, Donald Liebenberg's measurements showed that the intensity of this green spectral line oscillates strongly, with periods close to 6 min, as well as shorter periods. He believes he saw these oscillations repeat themselves several times, thanks to the duration of the totality offered by the Concorde flight. Such observation is difficult and has not been reproduced since this 1973 flight. Whether by plane or in space, Sun specialists prefer to directly measure the velocity of the coronal gas by analyzing the profile of a spectral line. If the gas is moving, the line is affected by the Doppler-Fizeau effect of the atoms which emit it and the velocity can be determined. We will come back to this in the second part of the book.

The Chromospheric Program of John Beckman

John Beckman, for his part, wants to determine more precisely whether these 20,000 °C of the chromosphere are uniform, or whether there is a mixture there with cold regions, coming from deeper layers where they are named "spicules." To measure the tiny infrared radiation of the chromospheric layer, John is not afraid to face the difficulties by using a bolometer operating at very low temperature, like ours. Its measurements are made one point of the image after another—today we would say they were made with a single pixel receiving the light. Just before totality, it therefore uses the movement of the edge of the Moon, revealing only a thin crescent and slowly cutting out the chromosphere. This cutting is slowed down since the plane flies almost at the speed of the lunar shadow, so that John, taking his time, measures very precisely the variations in the light intensity, and therefore potential variations in chromospheric temperature. The analysis of the measurements will have to take into account the lunar mountains which give the edge of the Moon a tormented profile, thus limiting the precision of the signals attributed to the different layers of the solar atmosphere. John's observation, made at far infrared wavelengths, was at that time unique, and it confirmed the mixing of hot and cooler elements in the chromosphere.

The Atmospheric Program of Paul Wraight

The Aberdeen team obtained excellent results on the emission from the mesosphere in the near infrared, between 1 and 2 µm wavelength. The ozone molecule O_3, made up of three oxygen atoms, is present in this atmospheric layer between 50 and 80 km in altitude, above the stratosphere. Ozone being dissociated into oxygen O_2 by solar ultraviolet radiation, this O_2 molecule gradually loses the energy then acquired by emitting two infrared spectral lines, which were successfully observed during the flight. These lines were well known from the observation of the luminescence of the night sky, nevertheless with many questions remaining about the characteristics of the emitting ozone layer, its altitude and its temperature. The very brutal change of ultraviolet solar light at the instant of totality but their excitation, much less sudden than during the eclipse, left. The g could provide useful answers to these. Indeed, the good results of Paul Wraight removed some of these uncertainties.[6]

6.3 We Were Not Alone

On board the Concorde, we knew of course that many astronomers, including Serge Koutchmy, had placed their observation devices in different points of African territory, whether in Mauritania, Niger or Chad. We also knew that, a few weeks before our flight, NASA had launched an ambitious laboratory into space, called Skylab, which orbited the Earth, well above the Earth's atmosphere at 420 km altitude, with a period of 93 min. Since May 23, three men were on board Skylab to carry out multiple scientific observations, including the study of the Sun and its corona with a Lyot coronagraph. At this altitude, the sky is black, the nearby corona therefore stands out, while the mask of the coronagraph hides the dazzling disk of the Sun. In addition, on June 30, Skylab's trajectory, which circled the Earth every 93 min, happened to come very close to the lunar shadow cone that swept the Earth. The result is an extraordinary photograph, where the new Moon, lit by the light of the Earth, appears on the background of the dark sky, a few moments before this lunar disk obscures the solar disk for observers who are located in the zone of totality. The corona is clearly visible, essentially the K corona crown with the jets of matter it ejects to space (Fig. 6.4).

With Skylab above us, we were not alone in the atmosphere either, as a four-engine NC-135 aircraft was flying 3000 m below the Concorde, with a team from Los Alamos on board, indeed the laboratory of Donald Liebenberg. Oddly enough, I do not remember Donald at any time mentioning to me this expedition from his own laboratory, which was flying below us, as evidenced by the astonishing photograph which will be published later (Figs. 6.5 and 6.6).

---

[6] Wraight and Gasden (1975).

Fig. 6.4 This picture is taken from Skylab, at the time it passes close to the band of totality. The Moon (upper left) is not aligned with the Sun, which is in the center, covered with a coronographic Lyot mask held by a support (shadow at lower left). The Moon is illuminated by the Earth's shine and the corona is well visible, enhanced by a numerical treatment of the image. (Source: NASA)

Fig. 6.5 Flying at subsonic speed, some 3000 m below Concorde 001, the crew of the NC-135 aircraft from Los Alamos Laboratory, took this picture of Concorde close to the third contact, when the light of the solar photosphere emerges and masks the faint corona. (Source: Los Alamos Laboratory)

Fig. 6.6 An artist vision of Concorde 001 during the totality. This painting, named Racing the Moon, was painted by the Canadian artist Don Connolly (Sydenham, Ontario) after a common elaboration with L. Robert Morris (Ottawa (Acrylic (49 × 68 cm), 2004. First Prize in the Commercial Category of the Aviation Week & Space Technology/American Society of Aviation Artists 2005 Aerospace Art Awards. Image assuming the viewer is located 3000 m below the Concorde flying at 17,602 m above Niger, at latitude 16.19°N, longitude 14.38°E, time 12:07:24 UT. The orientation of the jets in the corona is somewhat approximate, as shows a comparison with Fig. 1.1)). (Source: Don Connolly)

6.4 Was Concorde Spied Upon by a Flying Saucer?

It took the tragic destruction of the twin towers of the World Trade Center in New York on 11 September 2001 to show the citizens of the USA that their continent was not some kind of impregnable fortress that no one would ever be able to penetrate. No one, that is, except extraterrestrials! This fear comes from along ago[7]:

> On 30 October 1938, the CBS radio station in New York announced the live broadcast of a variety show given in one of the grand hotels. Five minutes later, the music was stopped to give a news flash, according to which the Mount Jennings observatory in Illinois had just observed curious and inexplicable vapor projections on the surface of the planet Mars. Then seismographs in New York recorded a tremor. A few minutes later, the somewhat shaky voice of the speaker announced that a large and mysterious object had just crashed into a factory in Princeton, New Jersey, and burst into flames. There were sounds of sirens screaming as the first emergency services arrived on the scene. Then, finally, a reporter began to give a 'live' commentary on the terrifying scenes taking place before his eyes. The Martians were disintegrating anyone who came anyone near the object …

The author of this radio hoax, which actually triggered scenes of panic in New York, was none other than a young actor by the name of Orson Welles, then only 23 years old, who made himself famous in the process. His aim had been to make a radio adaptation of *The War of the Worlds* by his close name-sake, the writer H.G. Wells.

After the end of the war in 1945, the Cold War stepped in to whip up a siege mentality in the USA.[8] The newspapers were full of reports of flying saucers, and even sometimes their crew. The US Air Force set up Project Blue Book, which ran from 1947 to 1969. During the McCarthy era, in 1952–1953, the US counter-intelligence agency (CIA) carried out its own inquiry, in great secrecy, of course. The US Congress was concerned about radio or any other signals that might tell potential invaders about the presence of inhabitants on Earth. France was not spared this strange fascination, with tabloid newspapers like *Paris Match* rushing out to report the slightest gleam of suspicious light in the sky. The government could not ignore public concern over this issue and the national space research institute (*Centre national d'études spatiales*, CNES) set up a special department to investigate unidentified objects spotted in the sky (*Groupe d'études et d'informations sur les phénomènes aérospatiaux non iden-tifiés*, Geipan),[9] which still exists in 2024. Pierre Guérin, an astronomer, belonging to the same research institute as Serge but a bit of an outsider, became the champion for the extraterrestrial origins of UFOs (unidentified

---

[7] http://www.forum-ovni-ufologie.com/t8611-la-guerre-des-mondes-les-martiens-d-orson-welles-le-30-octobre-1938#ixzz2a3iZLX5O.

[8] https://en.wikipedia.org/wiki/Unidentified_flying_object.

[9] https://www.cnes-geipan.fr.

flying objects), and the press made great use of his testimony to back up some of their less reliable claims.

Now, in one of the 36 Kodachrome slides of the African horizon taken by Jean at 12 h15 with his own 24 × 36 format camera, during the few brief moments when he had been able to contemplate the spectacle described above through a side porthole of 001, a suspicious dash of light appeared just above the horizon, against the perfectly black background of the stratosphere (Fig. 6.7). After enlarging the snapshot, this light became a tiny cloud with an apparent size close to that of the full Moon, white in the middle, orange-red toward the top, and fringed with green (Fig. 6.8). When the film was checked, it was shown that this was not due to a defect or scratch. Through an understandable lack of caution in Serge's lab, the press was informed in the days following our return and published the photo. There followed much speculation. Was it the Russians, whose Tupolev Tu-144 had so sadly crashed during a demonstration flight at the Paris Air Show in Le Bourget only a month earlier, who were spying on Concorde as it raced after the Moon's shadow? Or was it extraterrestrials, fascinated by this machine, a potential competitor for their own, who had also organized a rendezvous with a cosmic phenomenon that could be spotted from anywhere in the Solar System? And so on and so forth. What were these scientists hiding from us, given their lack of interest, even contempt, for UFOs?

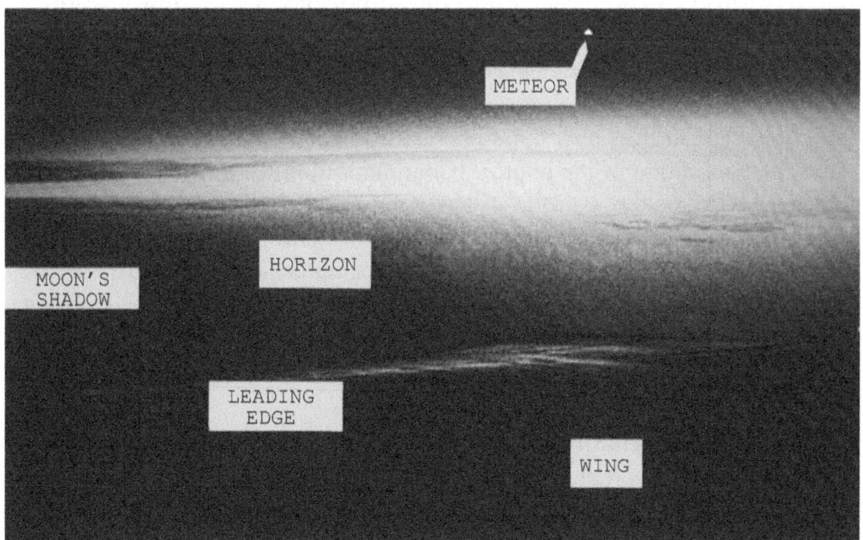

Fig. 6.7 Black and white photo taken from Concorde 001, showing a mysterious point of light. (Source: J. Bégot & S. Koutchmy/IAP-CNRS)

Fig. 6.8 Enlarged view of the object in Fig. 6.7. This image was used to prove that it was in fact a meteor. (Source: J. Bégot & S. Koutchmy/IAP-CNRS)

The national research agency (CNRS) was put in a difficult position and Serge Koutchmy had to give an explanation. To do this, with full support from Jean-Claude Pecker, head of his Institute and Professor at the Collège de France, he took his time and deployed all his many talents. In November 1973, he sent a brief statement to the *Agence France-Presse*, a Paris-based international news agency, detailed in a confidential and strictly scientific note, dated 8 April 1974 and passed on to the authorities. This is how it read:

The physical and geometric characteristics of the cloud show that the most likely hypothesis to explain this observation involves meteorite events of the "meteor eruption" type in the upper atmosphere. This hypothesis was revealed quickly after the publication of the photo". (Communication to Agence France-Presse dated 4.11.1973).

The following elements support this identification: (a) size and distance of the cloud, determined using geometric, photographic, and optical parameters; morphology of the appendix [of the spot of light]; (b) color suggesting emission of sodium D lines; (c) maximum visibility of the Beta Taurid [meteor] shower during totality [of the eclipse].

This UFO was therefore nothing other than an ordinary meteor of a few grams, burning up as it shot through the stratosphere at an altitude of some 50 km. Its green color, similar to the color of the aurora borealis, could be attributed to oxygen atoms present at this altitude, excited by their interaction with the passing object. It turned out that, at the date and time of the eclipse, the Earth happened to intersect the trajectory of some known meteorites as it went on its orbit around the Sun. These meteorites were invisible by day time and were just beginning to be known by their radio echoes on radars. Too bad for cheap sensationalism and so much the better for the scientific rigor of Serge Koutchmy! The management of Aérospatiale was not unhappy with this "affair," which had contributed to getting the Concorde plane talked about in many media outlets around the world!"

6.5 Afterthoughts

Strangely enough, I just cannot recall what triggered the idea of following an eclipse aboard Concorde, much as I would search my memory and even my notes. Using their telescopes, the Galileo I and Caravelle aircraft on which I had flown previously had brought me close to the stars. I vaguely recollect reading in a popular American astronomy magazine,[10] back in 1969, a suggestion by American colleagues who had sought to use the supersonic military aircraft SR-71 Blackbird, a spy plane, hence top secret, to follow the eclipse of 7 March 1970 on the east coast of the USA. But in 1970, the request for authorization was rejected by the US Department of Defense. If they had got permission in 1973, they would have been able to fly above us at an altitude of more than 20,000 m, thus obtaining two good hours of totality. However, they would only have had a tiny space in which to set up one instrument and only the two pilots would have witnessed the spectacular event. But they would have got the record! So maybe this little article had worked its way into my mind during the 2 years leading up to the lunch in Toulouse. I just don't know. The origins often remain shrouded in mystery.

There was no follow-up to the 1973 flight, despite our efforts in 1974 to try to repeat it with the British Concorde 002 prototype during the eclipse of 1976 above the Indian Ocean; with John Beckman, we had suggested a fascinating photographic program based on the excellent images obtained by Jean Bégot. But things are doubtless better that way, since our flight over Africa remains unique to this day. In any case, Air France didn't forget about it. When people began talking about the total eclipse of the Sun of 11 August 1999 that would cross France from east to west, the company decided to charter one of its Concordes, the FoxTrot Charlie F-BVFC, to take wealthy tourists to Ireland to admire the spectacle (Fig. 6.9). The intrepid astronomer Audouin Dollfus, then aged 75, was invited aboard.

---

[10] Mercier and Pasachoff (1969).

Fig. 6.9   Somewhere above the North Atlantic, from the cockpit of the Air France Concorde F-BVFC, specially chartered for a commercial tourist flight for a rendezvous with the Moon's shadow during the total eclipse of 11 August 1999, one that many in France would be able to observe. (Source: V. Coudé du Foresto)

Audouin was the son of Charles Dollfus, an aerostation pioneer whose little book *Histoire de l'aviation*, published on poor quality paper in the middle of World War II, had been one of my childhood favorites. Audouin had been a student of Bernard Lyot, the inventor of the coronagraph mentioned above, and had bravely made the first manned balloon flight into the stratosphere on French territory, reaching an altitude of 14,000 m. So here he was aboard FoxTrot Charlie, flown by the pilots Jean Prunin and Eric Célérier. With my young colleague and astronomer Vincent Coudé du Foresto, they delighted the passengers with details of the eclipse which turned out to be rather hard to see through the side portholes of the cabin. Totality was rather brief, a mere 5 min.

The music of Paul Verlaine still rings in my ears when I recall that beautiful morning in Las Palmas when we climbed aboard the great white bird that would carry us to the stars, in fact, toward our own star, the Sun, and our natural satellite, the Moon. The poet whispers[11]:

Dis, qu'as-tu fait, toi que voilà,
    De ta jeunesse?

While writing this story, I thought of all those who, in the prime of life, are already haunted by the poet's question and dream of their life to come. I don't

[11] *Tell me, what have you done, you there/With your youth?*

Fig. 6.10 Exactly 40 years after our absolute record for observation of a total eclipse of the Sun—74 min of darkness—eight members of that flight met at the Musée de l'air et de l'espace in Le Bourget on 29 June 2013, under the 001 prototype, in the presence of André Turcat (center). Left to right: Michel Rétif (test flight navigation mechanic) and the science team John Beckman, Donald Liebenberg, Alain Soufflot, Paul Wraight, Pierre Léna, Donald Hall

know how I would answer myself, if not by recounting my own story in these few pages, just as I experienced it with my companions in Las Palmas when Concorde 001 carried us up to the sky.

Fifty years after this flight, we know the solar corona much better, as shown in the second part of this book (Fig. 6.10). However, the corona has not yet revealed all its secrets. For a long time, we will still have to "catch a shadow as our only prey.[12]" It is with this beautiful expression that André Turcat described our historic flight, he who had wanted it, he who demonstrated that youth has not of age if not that of his dreams, he who left this world in 2016 at the age of 94, leaving us in a posthumous book this final and moving testimony of a thirst for learning throughout his whole life.[13]

---

[12] A play on the French expression "*lâcher la proie pour l'ombre.*"
[13] Turcat (2021).

Part II

Half a Century Later, Solar Corona and Exoplanets

7

The Corona and Its Gaseous Component

The solar corona is a large, more or less regular, luminous halo that surrounds the Sun, which is visible during a total eclipse of the Sun. As our understanding has developed, this halo has been gradually been interpreted as constituting the atmosphere of our star, extending far from it. Why does this atmosphere exist, what is it made of, how are its often very esthetic aspects and their variations over the years explained? These are all questions that have mobilized astronomers for several centuries, which have provoked many expeditions to observe eclipses, particularly that of 1973 aboard the *Concorde* 001 aircraft, recounted in the first part of this book. Since that date, what progress has been made, not only with more sophisticated instruments during the eclipses that have occurred all over the Earth, but also by placing in space, well above the Earth's atmosphere, telescopes that allow us to observe and measure the solar corona, even outside of eclipses and getting closer and closer to the Sun. Thanks to the use of the coronagraph, this brilliant invention of Bernard Lyot, these instruments have thus been able to observe from the ground the part of the corona closest to the Sun, and the brightest. Finally, the observed phenomena are better and better described thanks to the progress of the physics of ionized gases, called plasmas—that is, when the electrons of atoms are partially or totally torn from an atom by a high temperature, which occurs in the corona.

While difficult to study from the ground outside of an eclipse, because of its faintness compared to the solar disk, the solar corona deserves to be

The original version of the chapter has been revised. A correction to this chapter can be found at
https://doi.org/10.1007/978-3-031-92199-5_10

studied, particularly because of its importance in the relations between the Sun and the Earth. It remains a major subject of research, as more than 10,000 scientific articles have been devoted to it in the past half-century. It is obviously impossible to capture here the extent of all of these works. Their relevance goes beyond our solar system, since the dust grains contained in the corona and which constitute its extension into the zodiacal cloud are linked to the history of the formation of this system, five billion years ago.

The presence of these grains and their properties, as well as coronography, which is the experimental technique for studying them, today find a new field of application in the study of exoplanetary systems, and particularly the search for planets similar to ours, which we call exoEarths.

In this chapter and the following ones, we propose an illustrated walk through the twists and turns of the distended atmosphere of our star, the Sun, and then much further, toward these other Earths that astronomers are searching for and gradually discovering in this beginning of the twenty-first century.

In the first part of this work, the reader may have come away with the impression that there is no clear definition of the solar corona. A correct perception, as over the years, the understanding of this extremely tenuous but extended and dynamic part of the Sun's atmosphere has evolved. It was first identified as a halo with a mysterious composition, revealing itself during a total eclipse when the Moon completely obscures the solar disk. Subsequently, during the nineteenth and twentieth centuries, this halo, attributed to the Sun, revealed its complexity: a very hot envelope of gas, constantly blown outward from the Sun and reaching the Earth. Tiny grains of dust, much colder, mix with this envelope and are present in a zodiacal cloud, observed since antiquity, which extends beyond the Earth's orbit. We now examine the details of this envelope.

7.1   An Initial Overview of the Corona

During the few minutes of the splendid spectacle offered by a solar eclipse, everyone can discover with the naked eye this corona. It reveals itself, surrounding the disk of the Moon which during the brief minutes of totality hides that of the Sun and moves relative to the latter. This corona extends from a transition region, located just above the surface of the solar disk, which is then invisible, to the boundaries of the largest elongated structures, such as large jets or streamers and other plumes, which appear much larger than the Sun itself. The large coronal jets, also called streamers, which extend for millions of kilometers, are the most striking elongated structures that, in Assyrian or Egyptian accounts, suggested a winged Sun, whose traces are found in the bas-reliefs of Petropolis. More recently, during eclipses from the end of the nineteenth century, the observation of these stretched-out structures suggested jets of matter coming from the Sun, as can also be indicated by the polar plumes, which, during totality, stand out against the sky at the North and South Poles of the Sun. Thus, was born the idea of a "solar wind," a phenomenon now confirmed and important in the relations between the Sun and the Earth. These large structures have also led to the idea, now universally adopted, of a Sun showing the characteristics of a gigantic rotating magnet. Finally, the appearance of the corona and these structures changes, from eclipse to eclipse, with a periodicity of 11 years, called the solar cycle and which is found in other manifestations, such as the number of sunspots on the surface of the disk (Figs. 7.1 and 7.2).

Fig. 7.1   Solar corona, observed during the 2006 eclipse by Russian amateurs, operating in the Caucasus. Several hundred images were gathered and processed by computer to visualize simultaneously: **(a)** the K corona, a plasma with its structures stretched toward the interplanetary medium (solar wind); **(b)** the F corona of dust grains, which appears almost elliptical and more external. While showing the reddish character of the F radiation, which had already been discovered in 1973, the intensity of this dusty corona has nevertheless been greatly attenuated by digital processing, to let us see the structures of the K corona that overlap it. The image distinctly reveals many bright stars, which become visible during totality. The disk of the Moon is faintly lit during the eclipse by the "full Earth." (Source: Image processing by Aleksandre O. Yuferev [Novosibirsk] and contrast enhancement by Serge Koutchmy)

Fig. 7.2 Solar corona observed in Chile during the 2019 eclipse, during a minimum of solar activity, by N. Lefaudeux, French amateur astronomer and optical engineer. The image, taken during totality and showing several stars and planets, results from the digital processing of many individual images, which makes the F corona practically disappear. On the horizon, the low, distant layers of the Earth's atmosphere are out of the shadow and lit by the red light of the solar chromosphere, which the clouds scatter toward the observer. The large jets, spreading on each side in the Ecliptic, are spectacular. (Source: 2019 Total Solar Eclipse—HDR astrophotography, Nicolas Lefaudeux @ hdr-astrophotography.com)

This white halo, particularly intense near the edge ("limb") of the solar disk, is due to the scattering of the light from the solar disk, which illuminates free electrons coming from a gas essentially composed, as is the interior of the Sun, of hydrogen and helium, the atoms of this gas being dissociated (ionized) by the temperature of at least one million Kelvin that characterizes the corona. This white light is polarized, which results from the scattering phenomenon that is its source. The K corona is therefore the component of the global corona, formed by this cloud of ionized gas (called plasma) and emitting this white and polarized light. It is responsible for the permanent loss of matter from the Sun, which results from a quasi-radial outflow of gas toward space and whose origin is not yet completely understood. This quasi-radial flow of matter toward space is called the solar wind; it is made up essentially of protons and electrons that result from the ionization of hydrogen and helium atoms,[1] with, in far fewer numbers, ions of heavier elements, such as calcium, magnesium, iron, and many more.

The spectroscopic analysis of this white light, as well as the detection of radio waves emitted by the same region, established that this K corona, starting just above the solar surface, which is called photosphere then chromosphere as it rises higher, was formed of a hot gas, the temperature increasing with altitude to reach millions of degrees, without our understanding the reason for this abrupt transition. At such temperatures, the coronal plasma emits very high energy light—X-rays or extreme ultraviolet (EUV). It is impossible to observe it from the ground on Earth, as these wavelengths are completely absorbed by our atmosphere. However, from the 1970s, observatories from space were able to take images of the corona in X-rays and extreme ultraviolet. As the photosphere, which emits the white light of our Sun, is at a temperature of only about 6000 K, it is far too cool to emit these very energetic wavelengths, and as the telescope is above the Earth's atmosphere, a total eclipse is no longer necessary to observe the corona, and observations can follow the evolution of the corona from day to day, for months or years.

At a distance of about one solar radius above the surface of the Sun, a second component of the corona, called F corona, appears (Fig. 7.3). It is no longer formed of an ultra-hot gas. Its brightness is due to the scattering of the photospheric light by microscopic solid grains, mixed with the plasma but remaining at much lower temperature.

The hydrogen atoms ionize into a proton and an electron. The helium atom ionizes: either partially into a helium ion (He^+) and an electron; or totally into

---

[1] The hydrogen atom ionizes into one proton and one electron; the helium atom: either partially into one electron and one ion Helium+ or into two electrons and one ion, Helium ++, made of two protons and two neutrons, also called particle alpha.

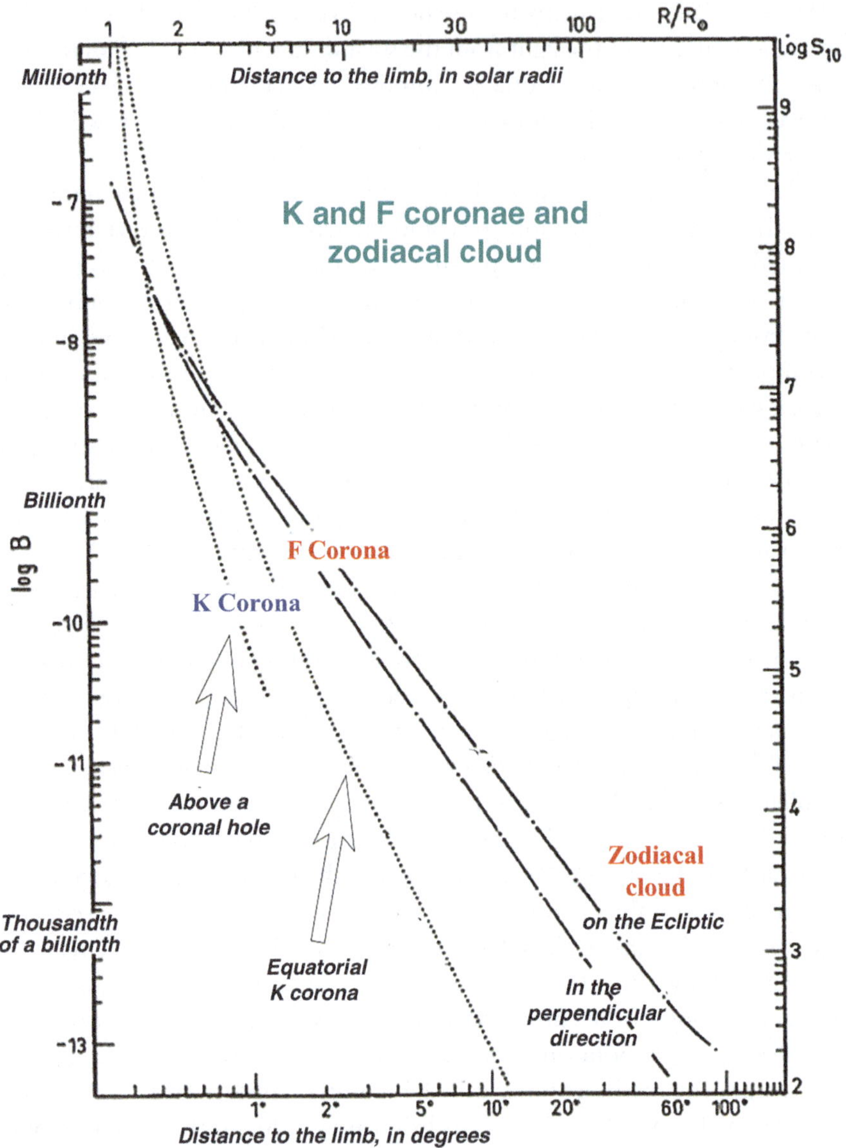

Fig. 7.3 Brightness of the K and F components of the solar corona, as a function of the distance above the edge of the Sun (solar limb), at the wavelengths of visible light. K corona, labeled in blue, with two dotted lines, either above a coronal hole or equatorial regions. F corona, labeled in red, extending further from the Sun as the zodiacal cloud, with two dot-dash lines either in the plane of the Ecliptic or in the direction of the solar poles. The brightness (called irradiance) is shown as a ratio to that of the surface of the Sun, on a logarithmic scale (that is, one millionth, ten-millionth, hundred-millionth, etc.). The distance to the center of the Sun is measured in degrees (bottom scale) or as a fraction of the radius of the Sun (top scale), knowing that the radius of the solar disk seen from the Earth is 0.25°. (Source: After S. Koutchmy & P. Lamy)

two electrons, and an ion made of two protons and two neutrons, noted He^{++}, also called an alpha particle.

In the same way as the electrons of the plasma, these grains also scatter the light coming from the surface, and their contribution gradually dominates that of the K component, the further the grains are far from the solar surface. The context of the study of this F component is quite different. It calls on celestial mechanics, accretion phenomena, collisions, and collisions between small solid bodies of the solar system, feeding this F corona with grains from new comets approaching the Sun, and it is finally linked to the origin of the solar system. What we learn from this F corona is valuable for exploring the exciting and new field of exoplanets, which is the subject of the last chapter of this book.

The intensity of visible or infrared light received from the corona in a given direction is therefore the sum of the light produced by the K and F components. The fact that the light from the K corona is polarized, while that produced by the F corona is practically not, allows them to be distinguished using an appropriate filter. Moreover, only the F corona emits significantly at infrared wavelengths. During an eclipse, it is observed that close to the edge of the Sun, the light from the K component dominates, while at a greater distance, beyond about two solar radii, the light from the F component dominates, forming the zodiacal cloud which is observable at night and outside of an eclipse up to 90° from the Sun and even beyond. Whether it is one component or the other, their brightness is extremely low compared to that of the solar disk, between one millionth (10^{-6}) and less than one thousandth of a billionth (10^{-12}). Even at high altitudes, this brightness is also very small compared to that of the "blue" sky, due to the scattering of sunlight by the molecules of the Earth's atmosphere: this is why the components K and F of the corona remain invisible when, outside of an eclipse, we look toward the Sun from the surface of the Earth.

The Sun is a source of a magnetic field, like the Earth but with a much stronger intensity. Our star behaves like a huge rotating magnet that fills the interior, the surface, and the corona with a magnetic field. Over the course of the 11-year cycle, this field increases in intensity then decreases, then the magnet field reverses, exchanging North and South Poles during the next cycle. This process has been most likely repeating for hundreds of millions of years. It was discovered by observing the periodicity of the number of sunspots appearing in the photosphere, and by that of the auroras on Earth, which are due to the flow of electrons from the Sun. To this general and relatively easy to represent magnetism, as produced by a simple bar magnet, is added the local production, on the surface and in the corona, of magnetic fields, variable over time and sometimes very intense in the photospheric spots. One could compare this to terrestrial cyclones, which locally accompany the more stable effects of the Earth's rotation on its atmosphere.

The appearance of the K corona during a total eclipse is very different depending on whether it occurs during a minimum or a maximum of the solar cycle. Understanding the role of this magnetic field on the structures of the K corona is therefore essential. As for the way in which the interior of the Sun produces this magnetic field and its periodic variation during the cycles, this is a difficult question, still quite poorly understood and into which we will not delve into here.

We will attempt to describe the physical properties of this bright and changing halo that is the corona, focusing on the properties within the reader's reach, and on points of general interest. However, many aspects, among which the mode of heating of the corona, capable of explaining its high temperature, as well as the eruptions and the acceleration of particles, the stretching, detachment, and migration of the magnetic field from the surface into the interplanetary medium and the solar wind, remain quite mysterious. From observations made with the coronagraph, attempts have been made, for example but without success, to explain the heating by the propagation and dissipation of waves (magneto–acoustic waves and the so-called Alfvén waves).

Heliophysics, the study of the Sun reaching out to the interstellar medium, has become a huge subject, addressing phenomena that occur in all layers of the Sun, from its core to its corona and beyond to the interstellar medium. It thus deals with the Sun's radiative energy, the transfer of radiation within the Sun, the convection of gas at all scales, the production of the magnetic field, activity cycles, acoustic vibrations (helioseismology), surface spots and faculae, filaments and prominences, plumes, flares (meaning a brief and intense burst of energy), and other eruptions. By successively examining the K corona, then the F corona, we are only going to share with our reader very modest elements of this heliophysics, which has become essential to understanding and predicting the multiple influences that the Sun exerts on our Earth.

7.2   An Extreme Richness of Details

Let us proceed step by step into the complexity of this solar corona and its two components K and F, as revealed by eclipses and which a simple diagram and some discussion allow us to better grasp.

Box 7.1 Components and Elements of the Solar Corona

K corona. First component of the solar corona. It is a gas mainly composed of hydrogen and helium, at very high temperature (million degrees) therefore ionized (the atoms are dissociated into nuclei and electrons within the gas). It extends to great distances from the Sun.

F corona.Second component of the solar corona. These are solid particles of matter (dust), mainly silicates, in the form of microscopic grains.

Lines of force of the magnetic field. They start from the solar surface, where the magnetic field is easily measured, and emerge away from the surface as shown schematically on the left of Fig. 7.4. They show the direction of the field at a point in space. Some of these lines form loops that close, others do not close and move away from the Sun into interplanetary space. They vary over the course of the 11-year solar cycle.

Coronal hole. Region of the lower corona where the value of the magnetic field is weak. Such a hole is represented on the figure at the North Pole of the Sun, the arrow indicating the rotation of the Sun on itself. Above a coronal hole, the lines of force of the field open toward space, which allows the coronal gas to escape: the local density is lower, hence the name "hole." The thin jets of matter form the "polar plumes."

Coronal mass ejection (CME). Ejection of gaseous matter (hydrogen and helium mainly), ionized (that is to say whose atoms are more or less dissociated into nuclei and electrons) toward space. This ionized matter carries away the magnetic field which is "frozen" in it—that is the magnetic field moves with the wind—thus of magnetic energy.

Flare (or burst). Local release of energy, resulting in an intense and transient emission of light (visible, but particularly ultraviolet and X-rays) within the hot coronal plasma. Noted "E" on Fig. 7.4.

Electrons of the coronal plasma. Noted e^-, these electrons result from the ionization of hydrogen and helium atoms produced by the very high temperature. Bound to the Sun by gravity, they are locally trapped by the magnetic field and scatter the visible light emitted by the solar surface, which produces the appearance of the K corona. This light is polarized.

Dust grains. Noted "g," these microscopic grains, most often formed of silicates (sand), are also bound to the Sun by gravity and also scatter the visible and infrared light emitted by the solar surface, which produces the appearance of the F corona. It is reasonable to assume that this light is not polarized, unlike the light of the corona K. The material of a cometary nucleus, when approaching the Sun, vaporizes. They form an ionic tail (effect of the solar wind on the gases from the nucleus) and a dusty tail (effect of the radiation pressure of the light on the grains from the nucleus) as shown on the right. This mechanism is probably partly at the origin of the grains of the F corona.

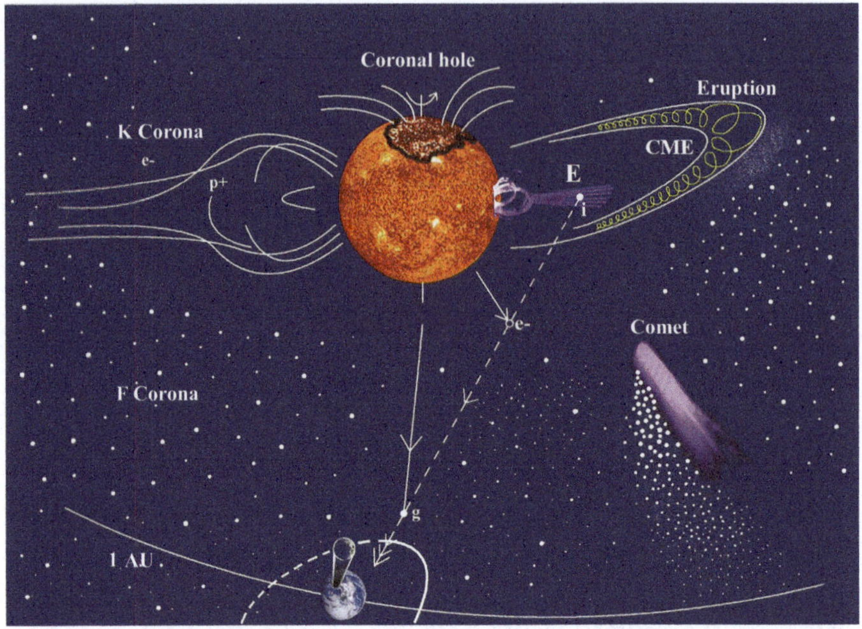

Fig. 7.4 Schematic representation of the solar corona, seen from Earth during a total eclipse. The diagram visualizes on the plane of the sky the different elements of the two components K and F, which contribute to the intensity of the light received at a point in the image. The solar disk is represented as it appears outside of an eclipse, when it is observed at the specific wavelength of a chromospheric spectral line

7.3   From the Sun's Surface to the Corona

There is little doubt but that the immense corona of the Sun is in one way or another connected with its surface and even with its deeper layers. Within these layers, an intense magnetic field strengthens or weakens, becomes complex in its organization, and dissipates its energy. It is the source and actor of solar activity, which manifests itself in an 11-year cycle of spots and many other phenomena, some of which directly affect the Earth.

To understand the corona somewhat, we need to present the foundation where it takes root, namely the surface of the Sun, which appears to us in the form of a bright disk. The luminous energy that the Sun radiates into space comes from its core, containing the nuclear furnace, very deep beneath the visible surface. However, it is from this surface, called the photosphere, that the gas density, and opacity, becomes low enough for light to freely escape into space. Most of the radiated luminous energy, received by the Earth, is found in the wavelengths of visible light, between the violet and red, which is a direct consequence of the temperature of the photosphere, about 6000 °K. Not only is this solar surface the main source of radiation escaping from the star, but it also forms the base of the corona whose magnetic properties are rooted there. Let us therefore present some elements of this surface, called the photosphere, and of the thin gas layer that surmounts it, called the chromosphere, forming a transition zone with the corona proper (Figs. 7.5 and 7.6).

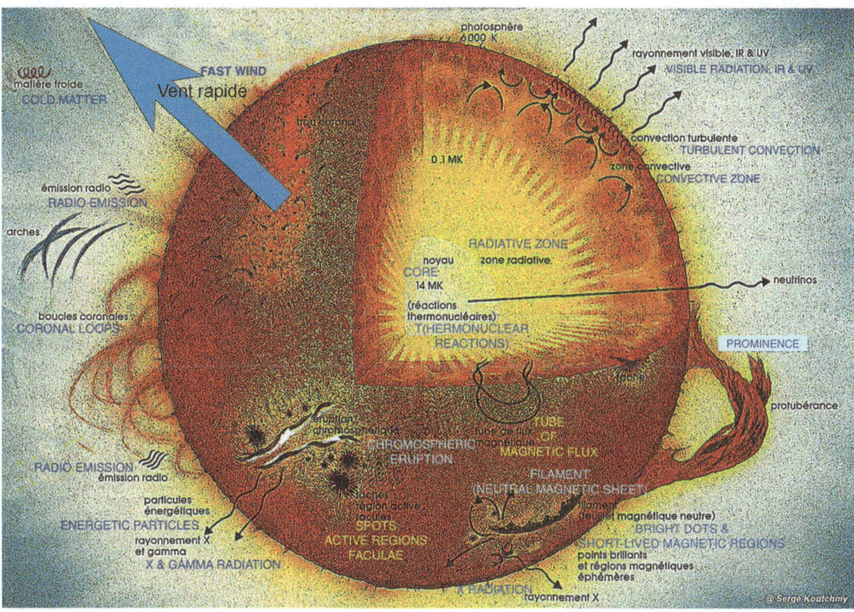

Fig. 7.5 The surface of the solar disk, with the different phenomena that determine the constituents of the K corona. The depths of the Sun are also presented in an exploded view, occupying the upper-right part of the solar sphere. Note: as original labels in French could not be removed, their English translation has been added. (S. K.)

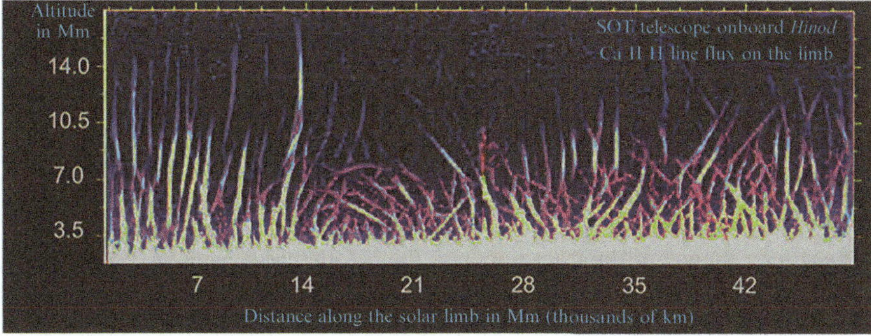

Fig. 7.6 Image of the spicule forest observed against the background of the sky, above the solar limb over a width of 15,000 km, in the chromosphere and at the base of the corona. Some spicules rise up to 15,000 km into the corona. The image is captured in visible light, within a spectral line of the calcium atom that has lost one of its electrons, using the 50-cm diameter telescope onboard the *Hinode* space mission in 2006, with collaboration between Japan, NASA, and Europe. (Source: NASA/JAXA)

Box 7.2 The Surface of the Sun

Chromosphere. Thin layer above the photosphere and forming the transition with the actual corona. The gas temperature begins to increase there (10,000 K), causing radiation from hydrogen and helium atoms, red in color (hence the name chromosphere), clearly visible albeit briefly during a total eclipse. The reason for this rise in temperature with altitude is not yet understood by heliophysics.

Convection. Vertical movement of a fluid heated from below when gravity (downwards) no longer compensates for the buoyancy (the force of Archimedes) upwards.

Turbulent convection. Appearance of random, disordered movements in a fluid, when the speed of vertical convection becomes too high. This phenomenon is easy to observe when the smoke from a cigarette rises in the air.

Magnetic energy. When a magnetic field is present in a plasma, where currents of electrons and nuclei (positively charged) also circulate, the energy associated with this field adds to the thermal energy of the plasma.

Chromospheric eruption. Intense release of magnetic energy in the chromosphere, particularly significant during a maximum of the solar cycle. This release of energy, produced by a sudden acceleration of the plasma particles, causes the emission of X-rays, then the ejection of matter outwards, which crosses the chromosphere and the corona, then reaches the Earth and its magnetosphere to produce polar auroras. Intense radio waves are also emitted, which also reach the Earth's ionosphere, disturbing it and affecting radio communications.

Facula. Warmer, therefore brighter, regions of the solar photosphere, while spots are cooler, therefore darker regions as their name suggests. The spots consist of a "shadow," as well as a "penumbra" of a different nature. The higher temperature of the faculae leads to more intense radiation, which compensates for the radiation deficit of the spots and can explain the variations in solar luminous flux during the 11-year cycle. The intense magnetic field of the faculae opens toward space and therefore contributes to the structure of the corona and the solar wind.

Filament (neutral magnetic sheet). Dense zone of the plasma, which is relatively cool (10,000 K), where the vertical component of the magnetic field is of zero intensity and where matter accumulates. Filaments appear dark on the disk.

Photosphere. Layer of the Sun, forming the visible surface of the disk, from which most of the luminous energy escapes into space in the form of visible radiation. The temperature of this layer is about 6000 K; it decreases to a minimum of 4300 K, located at about 500 km altitude, before increasing again.

Bright points and ephemeral magnetic regions. Very local and transient phenomena of magnetic energy accumulation, whose radiation in the extreme ultraviolet can vary very quickly (in a few seconds). These structures form a network over the entire surface, called the chromospheric network.

Prominence. Appearance of a filament when it is seen in the corona against the background of the sky.

Spicules. Thin structures, gas jets forming a kind of "grass" forest waving in the chromosphere, at the transition with the corona. Visually discovered on the solar limb during an eclipse as early as the nineteenth century by the astronomer Secchi, they are also observable outside of an eclipse with a coronagraph or in spectral lines emitted by the chromosphere. The matter in spicules is still relatively cool (10,000 K).

Sunspots, active regions, and faculae. Regions of the photosphere and chromosphere where the magnetic field can become very intense. For this reason, these regions are capable of supporting a higher pressure due to the coronal matter that overlies them. The corona above these regions consists of dense and hot loops, of varying dimensions.

Magnetic flux tube. The lines of force of the magnetic field appear as tubes, along which matter moves more easily than perpendicularly to the tube. In the tubes of the photosphere, the gas pressure decreases.

The figure also indicates the deep core of the Sun, where the thermonuclear fusion reactions release energy that propagates outward in the form of light radiation (radiative zone), then also, as it progresses outward, in the form of large-scale gas movements (convective zone), before being able to escape into space from the photosphere. Neutrinos, other particles than light photons, are also produced by the thermonuclear reactions of the core; they have the property of being able to escape directly from this core.

A beautiful illustration of the phenomena that occur in the transition zone between photosphere and corona is given by the observation of spicules, made in 2006 by a telescope in space. The detailed analysis of the agitation of these spicules shows the existence within the plasma of transverse waves, with a period of 100 s, with a velocity of several hundred kilometers per second. Such waves, called Alfvén waves, are known to exist in a plasma subject to a magnetic field. Here, they are complicated by a twisting movement, like in a tornado. These waves could contribute to the heating of the chromosphere.[2]

Box 7.3 The Transition Between the Photosphere and the K Corona

After this initial excursion, let us take a closer look at the transition between the photosphere and the corona, paying particular attention to the movements of the gas and energy exchanges within it, as depicted in Fig. 7.7. Indeed, this transition region plays a fundamental role in understanding the sudden increase in temperature, the transfer of ionized gas to the corona, and even the variations in the relative abundances of different chemical elements in the corona. At the temperature minimum (4300 K), located in the photosphere, there is a small bright region of concentration of the magnetic field B generated by local electric currents, with field lines represented by black arrows (the direction of the arrows is arbitrary). This small region is situated around a vortex or whirlpool. Convergent flows of gas (in green) with very little ionization, since the temperature is still low at this altitude, appear around 500 km above the visible surface of the disk. Organized large-scale convection movements form cells called supergranules, ubiquitous across the entire surface of the Sun, with a horizontal size of about 30,000 km. Too large to be represented in this diagram, they give way to the solar granulation on a smaller scale, which is also ubiquitous on the surface and

---

[2]Tavabi et al. (2011).

whose cells have an average size of 1000 km (or 1 Mm) (curved black arrows). In the dense layers of the surface, gas pressure is much higher than higher up, determining what happens above while expelling the concentrated magnetic field. In the slightly warmer chromosphere, there are ejections of cold gas or spicules, in red, with gas returning to the surface afterward (not shown), as well as more horizontal fibrils, aligned on a weaker magnetic field extending on the scale of supergranulation. Even higher, the X-ray and extreme ultraviolet radiation from the very hot but much thinner corona illuminate these lower layers. It is not easy to determine what this energy input implies for the thermal conduction of the gas.

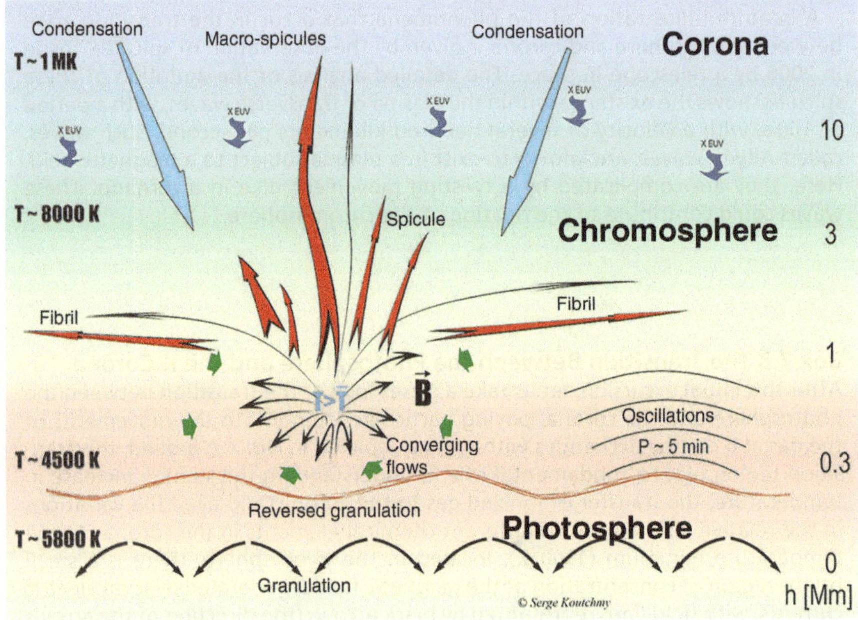

Fig. 7.7 Schematic cross-section of the chromosphere–corona transition region formed by the different layers located at the base of the K corona. The region is extremely heterogeneous in temperature (T: scale, in Kelvin (K), on the left), depending on the altitude (h: scale on the right). The right, logarithmic, scale, compresses the higher altitudes but allows one to look in more detail at the low altitudes. The phenomena of acoustic oscillations (bottom right) in the lower layers, as well as horizontal flows of matter, can induce magnetic waves or transverse disturbances of the quasi-radial force lines of emerging magnetic fields. These disturbances then propagate very quickly toward the corona, including in the form of Alfvén waves, or "kink waves," or even solitary oscillations, known as solitons, not represented here, which could participate in plasma heating

7.4   The K Corona and Its Structures

The brightness of the K corona, either in polarized white light or in its spectral emission lines and their polarization, has been measured during eclipses, allowing the examination of the inhomogeneities of the ionized gas (plasma) density and its temperature. The detailed distribution of the gas density can be directly deduced from the analysis of the fine structures of the corona. Alas, the resolution of the images is not sufficient to reach the very small scales, which are the most important for understanding the phenomena. The fundamental study of dynamic phenomena, rapidly evolving in time, is much more difficult during eclipses, given their brevity. We will nonetheless see an example later, during an eclipse in 1991.

For more than a century, increasingly sophisticated observation methods outside eclipse, have been used, both on the ground with Lyot coronagraphs and radio astronomy as well as from aircraft and in space. There, probes can measure far from the Sun and locally, by so-called in situ measurements, the properties of the coronal gas, such as density, temperature, and composition, while space observatories measure the X-ray and extreme ultraviolet (EUV) radiation of the corona. These probes use coronagraphs known as "external occulters," based on the principle of the Lyot coronagraph, but by placing a larger occulting disk in front of the telescope and not on the image of the Sun that the telescope forms inside at its focus. An additional internal mask improves the elimination of the residual stray light. Equipped with filters selecting a specific wavelength, they form images to which only this wavelength contributes if it is emitted by the gas.

What follows aims to give the reader a brief overview of what we know about coronal structures and physical phenomena within the K corona: its temperatures, densities, and speeds; the occurrence and structure of prominences and eruptive or stationary filaments, coronal holes, plasmoids, and ejecta, including spicules; its variability on large scales; its explosive phenomena—called CMEs, for coronal mass ejection (Figs. 7.8 and 7.9).

Understanding this corona is not simple, and many aspects of it are not yet satisfactorily grasped. Let us enumerate some of these questions: Why is its temperature so high and where does the energy that causes it come from? Are the particles constituting the solar wind ejected into space, on the one hand, and how is the corona fed from the transition region, on the other hand? Why does it change its appearance over the 11-year solar cycle, going from the state of calm corona to that of active corona? What role do the cool filaments and prominences in the hot corona, which suddenly appear as eruptions, play?

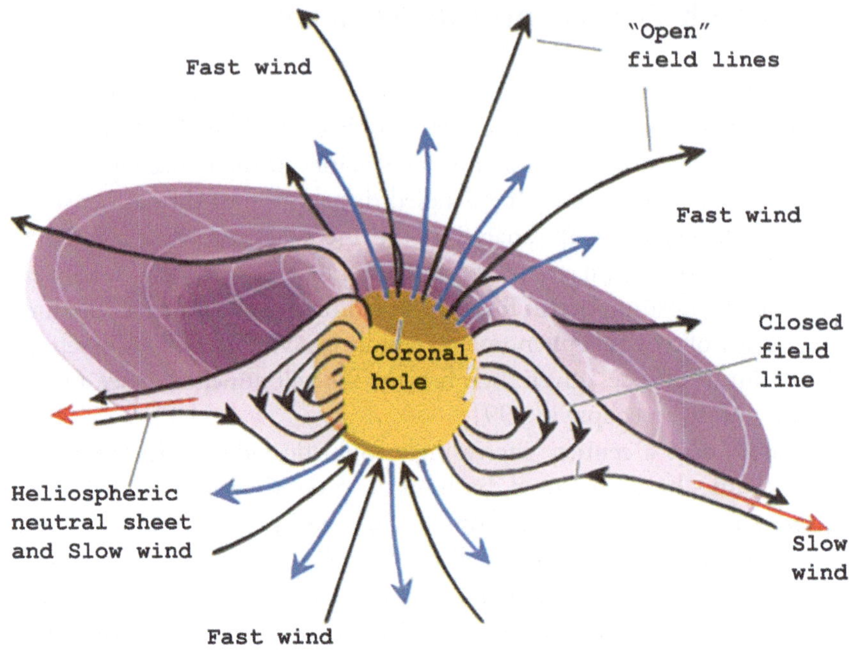

Fig. 7.8 The corona close to the surface of the Sun, during a period of minimum activity. In purple, a sheet-like structure, called the ballerina's or Alfvén's skirt. The lines of force of the coronal magnetic field (in black) trace out the skirt, with fields of opposite sense on each side of the sheet. The plasma is ejected in the form of fast solar wind from the coronal holes, and slow wind within the sheet. (Source: S. Koutchmy, adapted from S. Habbal and R. Woo)

What are the mechanisms producing flares (bursts of light) near the surface? These flares are mainly a source of energetic radiation (far-ultraviolet and X-ray photons) and a flux of energetic particles (protons, neutrons, and heavy nuclei whose energy reaches a million electron volts). Further into the corona, other flares only produce an emission of waves at radio frequencies (radio bursts). At the distance of the Earth's orbit, these particles pose a danger to all sensitive electronic systems, as well as to travelers in space, while the emitted radio waves can disrupt telecommunications.

The Magnetic Field and the K Corona

Since the K corona is formed of a hot and ionized gas called plasma, namely fast-moving particles of positive charge (protons) and negative charge (electrons), electric currents carried by these particles can form. Everyone knows

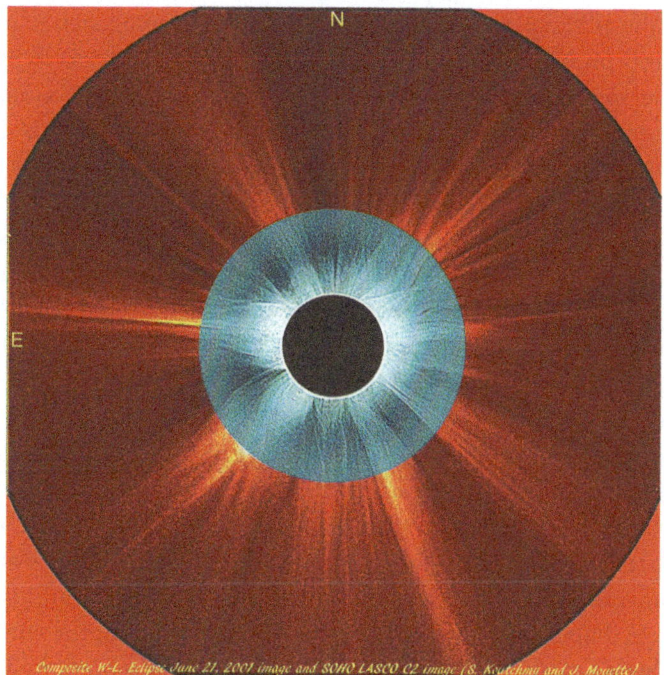

Fig. 7.9 Composite image of the K corona during the 2001 eclipse, observed simultaneously in Angola, for the inner and middle parts (in blue) and in space for the outer parts (in red), using the LASCO-C2 external occultation coronagraph of the European space mission SOHO. The perfect continuity of the large structures between the two parts of the image is striking. The Sun was then close to a maximum of activity during the cycle and the solar surface presented several new active regions of spots and faculae each month, while this number decreases or disappears at the minimum of activity. The component due to the F corona is subtracted by image processing and therefore does not appear here. (Source: S. K. & ESA)

that a magnetic field encountering an electric current exerts a force on it, called the magnetic force, which allows electric motors to work. The coronal plasma is therefore subject to this force, which will combine its effects with those, simpler, of gravity attracting particles toward the center of the Sun. Often, the magnetic force dominates over the force of gravity. In addition, an electric current being itself a source of a magnetic field, the effects of this second field on the plasma will add to those caused by the first. It is therefore not possible to understand these aspects of the corona without linking it to the magnetic activity of the Sun.

The intensity measured on the images in white light directly translates the plasma density in the corona, since the amount of light observed is directly related to the quantity of free electrons, and therefore to the total density of matter in the plasma, because it is made up of a homogeneous mixture of electrons and protons. As this mixture remains transparent to light, the terrestrial observer sees the contributions of all of the gas present along the line of sight. Examination of these images shows that the plasma is essentially confined in fine structures, that is to say filaments, loops, and sheets, as shown by images taken at total eclipses, in white light.

The intensity of the corona during an eclipse was first estimated simply by the naked eye, comparing it to the brightness of the neighboring stars, which became visible during totality. More precise detectors, simultaneously observing the corona and known stars, now allow a solid determination of the intensities of the K and F coronae and lead to the observation that the brightness of the F corona does not vary during the solar activity cycle.

The photometry of eclipse coronae was first carried out visually, by evaluating the light flux compared to those of the stars seen around the corona. Then photography allowed a better evaluation of the structures of the K corona, but the error bars were wide and only an analysis involving averages among the results from numerous observers allowed one to see real variations with the activity cycle. Finally, the precise photometry of the stars in the coronal field, observed during totality, allowed a more definitive determination of the absolute intensities of the K and F coronae, showing that the F component did not vary with the activity cycle.

Three beautiful images of the K corona, one global (Fig. 7.10) and the other two more local, visually illustrate, in their fine structures, these spectacular effects of the powerful forces associated with the magnetic field.

In the first image, strongly aligned fine structures appear. This suggests that they originate from a source in the lower corona that must be of a magnetic nature, and not just thermal, directed outward. The considerable energy carried by these thin beams of almost straight and directed particles over more than a million kilometers, as indicated by the arrow, thus indicates a source of energetic particles located in the lower corona. We have known for a long time a probably similar mechanism, accelerating beams of electrons at speeds close to that of light, and explaining certain solar emissions observed at radio frequencies (called Type III bursts).

This second image (Fig. 7.11), obtained at wavelengths in the extreme ultraviolet, in space and outside of an eclipse, shows fine details, measuring 400 km on the Sun and thus giving the thickness of a loop. These loops of ionized gas are confined by an invisible magnetic field, and the gas can flow toward the surface, forming "coronal rain." Gas, either cooler or8 hotter, is also present, but does not emit X-rays at this wavelength.

Fig. 7.10 This global image of a very extended K corona results from the composition of several images and digital processing. It shows very fine structures, stretched in the external K corona, clearly extending structures of the inner corona. It brings together an image of the 1999 eclipse (blue part), and the external structures (red part) observed simultaneously and outside of the eclipse, from space, with the LASCO C2 coronagraph of the European SOHO mission (*Solar and Heliospheric Observatory*) (Koutchmy et al. 1978). *Note: the white spot on top of image is a processing defect, which unfortunately could not be suppressed.* (Source: S.K. & ESA)

The cross-section of a loop does not seem to vary much with altitude, which distinguishes it from a magnetic tube, where the flux of the magnetic field remains constant.

The third image (Fig. 7.12) also shows the fine details of these structures, exceptionally observed at a very small scale, during the eclipse of July 11, 1991. Indeed, the narrow zone of totality of this eclipse occurred over the Pacific Ocean, and swept the large *CFHT* (*Canada-France-Hawaii Telescope*) of 3.6-m aperture, located on Mauna Kea on the Big Island of Hawai'i. Thanks to the resolving power of the details enabled by the large diameter of this telescope's mirror, these images have shed new light on the study of small-scale phenomena, where gas turbulence occurs and where magnetic energy heats the gas.

Fig. 7.11 Detail of coronal loops, forming above a chromospheric active region presenting a burst or flare, observed outside of an eclipse in extreme ultraviolet (iron IX and XI lines, corresponding to a temperature of one million Kelvin), from space by the TRACE mission (Source: NASA)

The effect of the magnetic field is strongly suggested by the presence of coronal loops. An inextricable tangle of loops is revealed. The interpretation of these loops shows the importance of the magnetic field, which can be deduced from the electric currents located at the surface of the Sun and in the chromosphere–corona transition region. Nevertheless, on the left of the image, a more turbulent part of the corona appears. Finally, radial and aligned, long and thin structures are superimposed, appearing in projection on the plane of the sky. Small structures called plasmoids, evolving during the brief duration of the eclipse and whose existence had been predicted by theory, have also been observed, probably in connection with the solar wind.

We have already emphasized that the photosphere and chromosphere, being too cool, do not emit light at X-ray and far-ultraviolet wavelengths. The observation of the solar disk from space at these wavelengths therefore shows a "face-on" corona for the terrestrial observer. Let us conclude this correspondence between structures seen at different altitudes, and here face-on (outside of eclipse and from space) or in profile (from the ground, during an eclipse), with two composite images of the K corona.

Fig. 7.12 Coronal loops. Detail of the composite image of the solar corona, obtained with the 3.6-m *Canada-France-Hawaii* telescope on Mauna Kea, in Hawai'i, during the total eclipse of 1991, at a maximum of the activity cycle, with a spectral filter transmitting visible light, but excluding any emission spectral lines (November and Koutchmy 1996). The original image underwent digital processing with the Mad Max (Tavabi et al. 2013) algorithm. (Source: adapted from Koutchmy & November, with permission of AAS)

In this first image (Fig. 7.13), the polar regions show good correspondence between jets or plumes, linked to small structures near the solar limb. However, there is little correspondence at low latitudes around the solar equator. Several reasons can be invoked, such as the effect of higher temperatures in more active regions, or the superposition of loops along the line of sight.

This second image (Fig. 7.14) corresponds to the corona during a period of minimum solar activity. The image combines an observation of the corona in visible light during an eclipse, showing very beautiful jet structures, with an almost simultaneous observation made from space in the extreme ultraviolet, where the face-on corona stands out on the disk (Fig. 7.15).

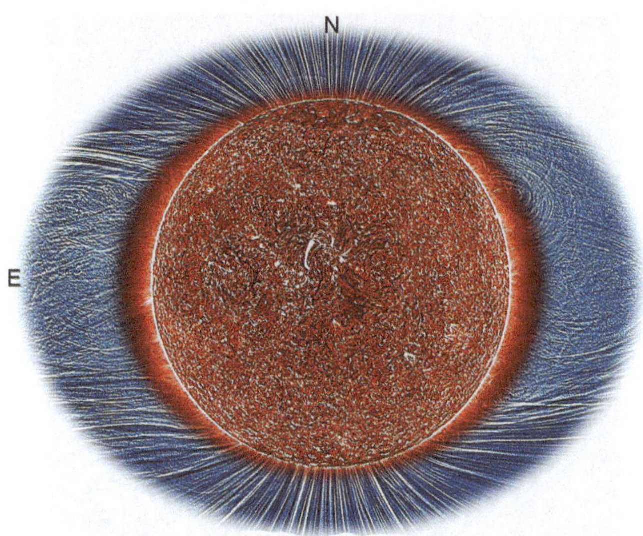

Fig. 7.13 Two views of the K corona are gathered here to highlight the continuity of structures. On the solar disk, the corona (in red) is observed from space and outside of eclipse, at wavelengths of the extreme ultraviolet emitted by iron IX atoms (ionized eight times) and iron XI, whose spectral lines are sensitive to a coronal temperature of one million Kelvin or less. (From the AIA instrument of the *Solar Dynamics Observatory*, launched in 2010). Outside the disk (in blue), the white light image of the corona during the 2019 eclipse (Fig. 7.1), synthetic image of N. Lefaudeux at the same scale and with the same orientation (Source: S. K. & N. Lefaudeux)

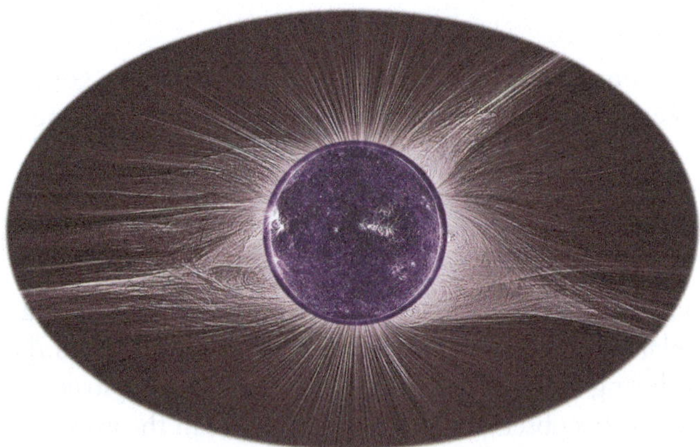

Fig. 7.14 Oriented image of the corona, during the total eclipse of 2017, observed by Nicolas Lefaudeux in the United States (a French amateur with an excellent site: https://hdr-astrophotography.com/) in white light, close to the minimum of solar activity during the cycle, and, almost simultaneously, with the camera of the *Solar Dynamics Observatory*, launched in 2010 (NASA). The disk of the Moon has been replaced here by an image of the face-on corona on the disk, made at the wavelength of 17.1 nm in the extreme ultraviolet. (Source: Processing by N. Lefaudeux & composite by S. Koutchmy)

Fig. 7.15 Magnetic field line structures. This aspect of the field is calculated using a simple axisymmetric model of two electric current loops, of opposite direction, whose radii are in a ratio of ten, flowing near the surface, in order to simulate very approximately a sunspot and its surrounding faculae of opposite polarity, following an old suggestion by physicists Lev Landau and Evgueni Lifchitz in their physics course considering the hydrodynamics of continuous media. More realistic variants can be easily calculated by changing the radius ratio and especially by offsetting the axes of the loops. The intensity or amplitude of the magnetic field appears in shades of greenish gray, superimposed on the field lines in solid lines. The black region located above the spot or first current loop, around the singular point where the field cancels out, forms a presumed magnetic trap where turbulence could dissipate and consequently heat the plasma. One can also note the resemblance of the force lines of this simulation with the structures observed in the corona during eclipses. Moreover, plasma-wave fluxes are well observed above the shadow of sunspots. Finally, above the singular point a neutral region extends radially (in reality, a neutral sheet) where the plasma is confined as in the large coronal jets. (Source: Numerical model by S. Koutchmy, L. Soloviev, M. Molodensky et al., *Space Science Reviews* 70, 283–285, 1994)

Box 7.4 The Forces Within the Coronal Plasma

The more intense the magnetic field, the more it exerts a force on the charged particles of the plasma. Next to the classic, gas pressure that characterizes (product of density and temperature), there is therefore here a magnetic pressure. Its effect is to align the particles along the lines of force of the magnetic field, which connect the North Pole to the South Pole of the solar magnet. The classic pressure is stronger when the gas is dense and its temperature is high. We often characterize the plasma by the ratio, called β (beta), between the local energy related to the agitation of the gas particles (classic pressure) and the energy contained in the magnetic field (magnetic pressure related to the square of the magnetic field's amplitude). The curvature of the field lines, the presence of turbulence and explosive events can complicate the analysis. If the factor β is small, the movements of the plasma in the corona are along the magnetic field lines. Let us recall that the magnetic field vector at a point results from the sum of the induced field originating from the surface of the Sun and the field created by the currents of the coronal plasma.

If we ignore for a moment the magnetic pressure (β large), gravity, directed toward the center of the Sun, opposes the pressure of the very hot gas, which gives a considerable extension to the coronal atmosphere due to the high temperature. Height scales are easily calculated and extend over more than one solar radius (the scale height is about 1/10 of a solar radius, near the surface, and grows with height due to the increase in temperature and the decrease in solar gravity). However, it is clear that the corona is not organized, like the Earth's atmosphere or the depths of the Sun, into strictly successive spherical layers controlled by only the force of gravity, not to mention that these layers are the site of dynamic phenomena.

The lines traced by the magnetic field high in the corona are of two categories: open or closed. The open lines start from the solar surface and are lost far into space in an Archimedean spiral. Near the Sun, X-ray images show that they start from the surface from darker regions, called coronal holes. Magnetograms show a dominant polarity in these regions. The closed lines, forming arcs or loops, are clearly visible around the active regions of the surface (spots and faculae) or around the chromospheric filaments that move away from them. The violent and sudden ejections of coronal matter (CME) can even "carry" with them a magnetic field. Thus, on white light photographs taken during an eclipse above the limb of the Sun, feather-like or polar jet structures corresponding to coronal hole regions appear open, just like the ambient magnetic field lines. Conversely, arched and looped structures are clearly visible around active regions and chromosphere filaments corresponding to closed magnetic field lines that are strongly disturbed or even detached during a CME.

This description of these movements of a plasma subjected to a magnetic field is called the MHD (magnetohydrodynamic) approximation. It can explain the forms, of great beauty, observed in the images and films of the K corona with the matter "frozen" to the magnetic field lines obeying the elementary laws of Gauss, Ampère, and Faraday.

It is not easy to measure the electric currents in the plasma remotely, currents that are also a source of magnetic field. It is possible to describe the magnetic field induced in the corona by the most powerful currents, those that are due to the motion of ionized matter within the dense layers of the photosphere, underlying the corona.

With the help of a simple model, it is thus possible to estimate the induced magnetic field above an active photospheric region, composed simply of a spot (cool) and a facula (hot). A double ring of coplanar photospheric currents, which we know to be quite realistic, allows the precise calculation of the magnetic field. The magnetic configuration that results from this calculation strongly resembles the typical structures observed in the corona.

A small field therefore has no effect (β very large). This region of small magnetic field can therefore be considered as an area capable of accumulating coronal plasma. This is a simple model that highlights the existence of gas concentration surfaces, as well as particular points (called singular or neutral), where charged particles can change trajectories and thus modify the magnetic configuration and simplify it.

Thus, in certain circumstances (not too turbulent plasma), when the opposite magnetic field polarities cancel out, the effect of gas pressure increases abruptly, causing rapid transient phenomena that raise the temperature (heating). Other models describe events in terms of so-called magnetic reconnection.

Box 7.5 Numerical Models of the Coronal Magnetic Field

Box 7.4 has proposed an example, where the magnetic field in the corona is calculated from a relatively simple configuration of electric currents in the underlying photosphere. Would it be possible to apply a similar method from more elaborate configurations of photospheric currents, ignoring the sources of magnetic field that are the currents within the coronal plasma? Thus, from basic physical laws and by a numerical calculation, an attempt is made to develop a representation, as faithful as possible, of what the real corona might be (this approach is called numerical modeling).

The ultimate goal of a simulation, such as the one presented in Fig. 7.16, would be to predict large-scale instabilities, which could explain the start of a CME event.

The overall morphology of the corona resulting from such calculations is comparable to the density structures observed in the images. However, regarding the small-scale and highly detailed features that are abundant in the corona, such as jets and rapidly changing filaments, the model does not provide satisfactory simulations. For example, it does not predict the existence of plasmoids[3] (sometimes referred to as "blobs"), which are observed in the solar wind as small clouds within which the magnetic field is closed.

A second stage of simulation must take into account the electric currents in the corona, and the resulting magnetic forces, as well as any sudden changes in magnetic configuration, called discontinuities with their own instabilities. This describes the beautiful coronal loops, whose foot points are anchored in the surface.

However, care must be taken in the calculation to ensure that some field lines do not close back on the Sun, but rather continue into space beyond a solar radius from the surface, as observations clearly indicate. This can be achieved by better accounting for heating and coronal temperature in the model, which appears higher the longer the loops are (scale law considering the loops as isothermal.)

---

[3] Delannée et al. (1998), Zhukov et al. (2000).

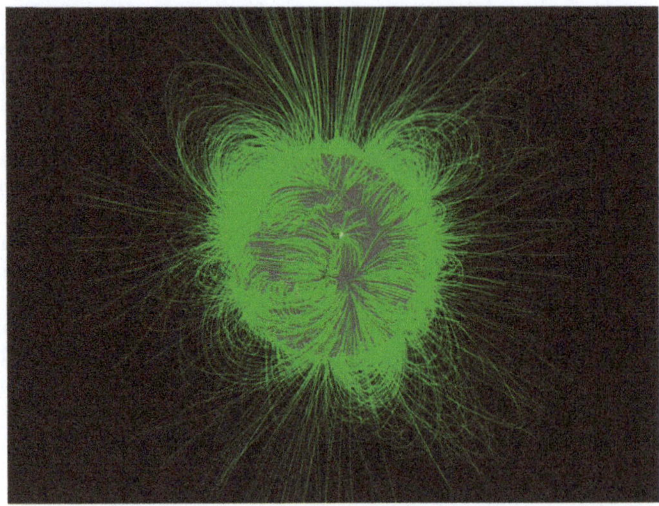

Fig. 7.16 Model of magnetic field lines calculated around the Sun in 3D and projected onto the plane of the sky. The calculations are made with the assumption of a current-free field in the corona. In addition, and in order to force the field lines to open outward and let the plasma forming the solar wind flow away from the Sun—according to what is observed-, one artificially introduces a source of magnetic field (electric current) at about one solar radius from the surface. It is further assumed that the corona is stationary during a full rotation of the Sun (no CMEs or dynamic phenomena on short time scales). To go further and make the simulation realistic, the equations, which describe the three-dimensional distribution of densities and temperatures, must be treated numerically with powerful computers. Ultimately, heating sources are artificially introduced into the calculations to open the magnetic field lines and let the plasma flow into the interplanetary medium in the form of solar wind. (Source: NASA, https://svs.gsfc.nasa.gov/3286)

Let us conclude with two examples this incursion on the role that the magnetic field plays in the acceleration of particles and the release of intense amounts of light energy.

The events called CMEs, which are the main manifestation of coronal activity, are ejections of matter, thus of coronal mass. Sometimes, they have been observed during the brief minutes of total eclipses. These ejections are caused by "magnetic explosions" in the inner corona, within large-scale structures that become, more or less suddenly, unstable. For obvious reasons, space coronagraphs, with their very long duration time coverage, are much better suited to observing such events from an orbit around the Earth or even better, from an orbit about the Sun, for example with NASA's *Solar Terrestrial Relations Observatory* (STEREO) mission launched in 2006.

Similarly, we illustrate with a pair of images, giving a view in relief (stereoscopic), the heating caused by a sudden release of magnetic energy in the coronal gas, above an active region located on the disk. These two images are taken at wavelengths of the extreme ultraviolet (EUV) in a region of the spectrum centered on a line of ionized iron, that of Fe XV, at a wavelength of 28.4 nm. This spectral line is sensitive to high coronal temperatures, around 2–3 million degrees Kelvin, and its presence results from the thermal excitation of ionized atoms, thus illustrating the heating related to the underlying magnetic field (Fig. 7.17).

Spectroscopy, a Tool for Exploring the K Corona

The intensity of light emission by a gas, depending on the wavelength of this light, is called the emission spectrum of this gas. The spectrum can be continuous, it contains all the wavelengths, at least in a wide range, as is the case for the light of the Sun that illuminates us, of the K corona or of the light of the F corona. The spectrum can also be strongly discontinuous, when only certain wavelengths, specific to atoms or ions of the gas and called "spectral emission lines," are radiated. This is the case of the coronal lines observed in extreme ultraviolet and in X-rays. Finally, a continuous spectrum can also "lack" certain specific wavelengths, which results in black lines in this spectrum, called "absorption spectral lines." This is the case of the photosphere spectrum in visible light (Fraunhofer lines). As it is precisely this light that the grains of the F corona scatter, the radiation from it also has black lines.

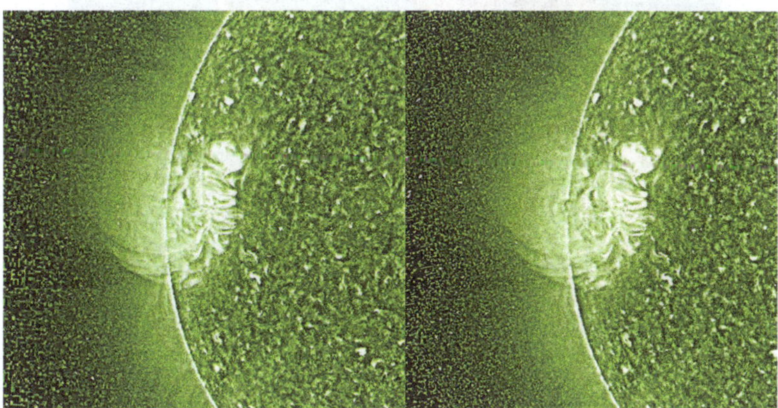

Fig. 7.17 Pair of stereoscopic images made using images taken simultaneously by the two NASA STEREO interplanetary probes traveling around the Sun on heliocentric orbits and located ahead of and behind the Earth, on March 22, 2007. The angle between the viewing directions of the Sun, from the two probes, is 2.5°. [To see "in 3 dimensions," merge the two images by placing the eyes so as to see the two images overlap and converge your eyes on a finger placed halfway.] (Source: STEREO/NASA)

Several images of the corona or chromosphere have been presented above, with the indication that these images were made with a filter, which isolated a particular wavelength of the light, whether it was visible, ultraviolet, or X-rays, rejecting any other radiation. These images can be qualified as monochromatic, even if strictly the filter does not let pass a single wavelength, but a narrow range around this wavelength. We then pointed out that the intensity of the light in the image was a direct indication of the temperature of the emitting gas, and that it was possible to choose the filter to select this or that temperature range in the exploration of the corona. This series of images during the solar cycle (Fig. 7.18) is another spectacular illustration.

At the wavelengths of visible light, the white light of the K corona is mainly due to the scattering by the electrons of the coronal plasma of the light coming from the photospheric surface. Its spectrum is therefore continuous, containing all the wavelengths of the rainbow. The black lines disappear during the scattering phenomenon, due to the very high thermal speed of the electrons which, by the Doppler–Fizeau effect, "scrambles" the spectrum of the scattered light. Like the lower layers of the Sun, the corona also contains, in very small proportion and heated to very high temperature, atoms and ions of

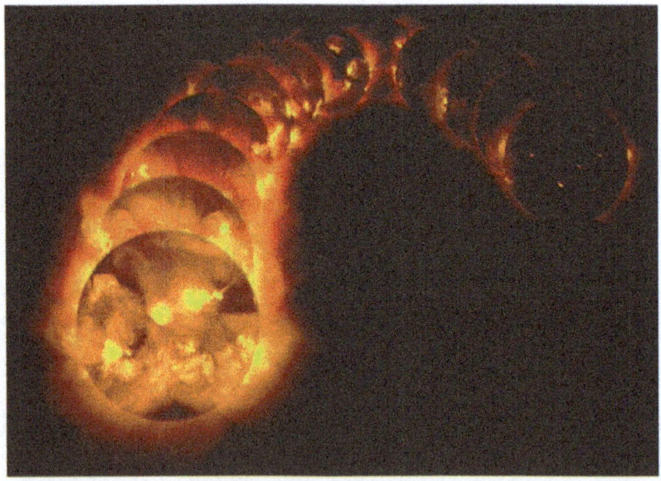

Fig. 7.18   This montage brings together images of the corona, taken from space at the wavelengths of spectral lines in X-rays, for more than 10 years (*Yohkoh* mission of the Japan Aerospace Exploration Agency JAXA & NASA). The temperature of the bright areas reaches two million Kelvin. During the activity cycle, the intensity of the lines is very intense at solar maximum (on the left) and almost nil at solar minimum (on the right). This Japanese mission collected more than one million images between 1991 and 2001 before falling victim to a solar eclipse! The darkness in which the satellite was plunged affected its power supply and disrupted its attitude control, which became impossible to restore. (Source: Lockheed-Martin Solar and Astrophysics Lab)

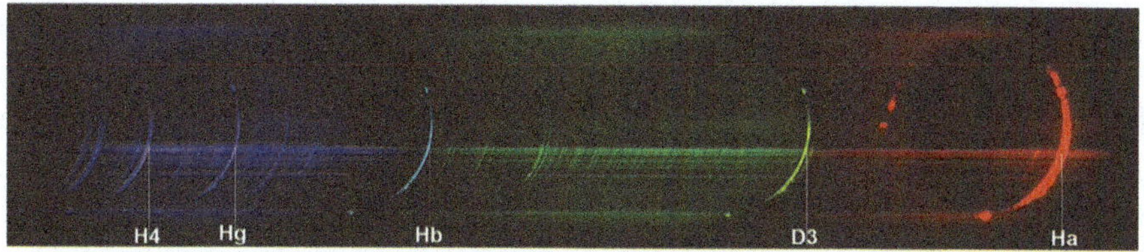

Fig. 7.19 Flash spectrum of the chromosphere. For a few seconds, before and after totality (hence the name "flash"), the disk of the Sun is totally obscured, except for the thin layer of thickness about 5000 km (about the radius of the Earth), located between the limb of the Sun and its corona. It is possible then to photograph its spectrum. The spectral lines are those emitted by the chromosphere, where the temperature of the hydrogen and helium gas remains low (less than 10,000 K). Un-ionized (neutral) hydrogen emits a strong red spectral line, called Hα. We see another, orange, line called D3, due to neutral (un-ionized) helium, to the right of the spectrum, as well as lines in the green–blue. All together, they give a pink color to the chromosphere. (Source: an anonymous amateur astronomer)

almost all the chemical elements, notably oxygen, carbon, calcium, nickel, and iron. Some of these atoms resist the temperature better and do not lose all of their electrons.

As with any atom, the interaction with the light of such an atom, strongly ionized, is governed by quantum physics and leads to an ability to re-emit or re-scatter light at one or more specific wavelengths, strictly specific to the considered chemical element. A spectrum of the emitted light reveals spectral lines at these wavelengths, and no light at others (emission spectrum). In addition to the possibility thus offered to characterize the emitting element, the spectral lines allow the determination of other quantities, specific to the atom's situation: temperature of the gas it belongs to, but also its movement speed, local magnetic field, etc. These lines, observed with the instrument called spectrograph, are therefore extremely valuable for knowing the environment containing the atoms. In particular, it is possible to make images only at the wavelength of a chosen line and thus being sensitive to the value of a particular parameter, for example the temperature.

The beautiful image (Fig. 7.19) of the chromosphere's spectrum clearly shows the emission lines at the Sun's edg, which dominate very largely over the continuous spectrum.

Spectral analysis is an extraordinarily powerful tool for astrophysics, and here for heliophysics.[4]

---

[4] For a modern presentation, refer to the so-called CHIANTI numerical code developed by an international consortium and kept alive by del Zanna, et al.) (International-consortium 2020).

One of the most famous discoveries due to spectroscopy is that of the chemical element helium in 1868. The French astronomer Jules Janssen detected a hitherto unknown line during a total eclipse observed in Guntur (India). Its presence was confirmed a few months later, outside of an eclipse, by the British astronomer Norman Lockyer, and the confusion with a possible line of the sodium element was lifted. A new chemical element was discovered, which Lockyer named helium, from the Greek name for the Sun. A few years later, its existence on Earth was confirmed by the spectroscopy of lava from Mount Vesuvius by the Italian volcanologist Luigi Palmieri.

Considered audacious in the 1930s, the hypothesis of a very hot corona was indirectly supported by certain observations: the immense radial extension of the corona, or the morphology of the large jets stretching outwards and never similar from one eclipse to another, which suggested permanent gas flows escaping. Until 1940, astronomers observed spectral lines in the corona during eclipses including a fairly intense green one, which was impossible to attribute to a known chemical element, because as in the case of helium in the previous century, the line had never been observed in the laboratory. The Swede Bengt Edlén and the German Walter Grotrian identified the origin of these lines that are said to be "forbidden," because they are not emitted in normal conditions but only under the physical conditions particular to a very low-density gas where collisions between ions and electrons become very infrequent. They were emitted by iron atoms heated to a million degrees, therefore highly ionized since they had lost half of their 26 electrons.

Subsequently, other elements were identified in the spectra of the corona, such as calcium or nickel, all highly ionized. Later, the detailed measurement of the profiles of these lines, with Bernard Lyot's coronagraph, indicated for the first time that the emitting atoms (ions) had agitation speeds characteristic of a gas at this same high temperature—a million degrees. Thus, immediately above the chromosphere with its forest of spicules, the temperature of the corona rises to exceed one million degrees.

For several decades now, access to space has significantly advanced work on the problem by allowing the study of the emissions allowed from a large number of other lines located in the extreme ultraviolet (EUV) and X-ray domains. As already indicated above, the much cooler surface of the solar disk (photosphere) does not emit radiation in these domains: the corona therefore stands out against this dark background of the cool disk. With their different sensitivities, these lines allow the exploration of a wide range of temperatures, ranging from 50,000 to 5 million degrees Kelvin and even much more when it comes to ephemeral flares where a huge amount of energy is released that heats the gas. The possibility of choosing filters corresponding to spectral lines sensitive to this or that temperature zone leads to very complete observations, as is the case of the NASA *Solar Dynamics Observatory* mission, for example. This observatory collects every 12 s, and continuously, a group of 16 images, each in a format of 4000 × 4000 pixels. This provides information far superior to that obtained so far during eclipses and, moreover, continuously. The following images highlight this remarkable exploration tool that is the spectroscopy of the corona.

Figure 7.20 illustrates the wealth of information that an image obtained at a single well-chosen wavelength provides. Huge dark filaments cover the surface at this altitude and are manifested outside the disk as prominences, for example, in the Northeast (upper-left). The bright areas of the disk

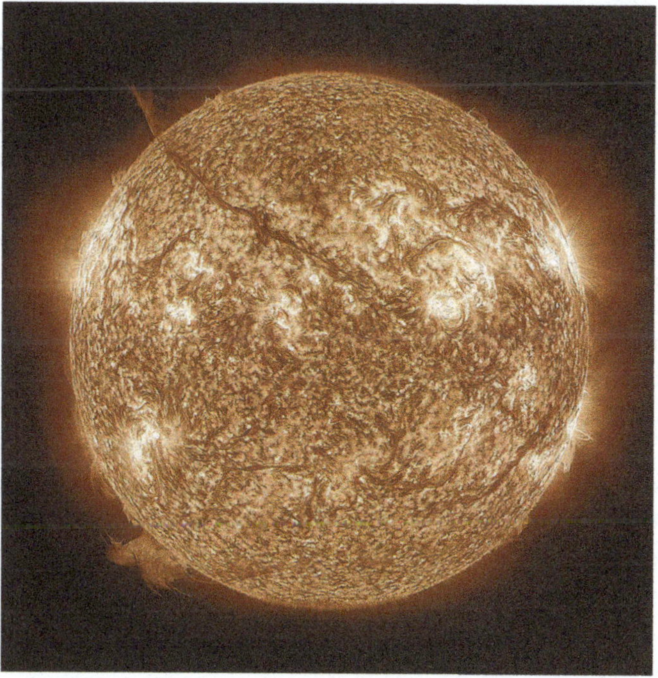

Fig. 7.20 The Sun observed in the resonance line of ionized helium II at 30.4 nm (extreme ultraviolet). This line, emitted at a temperature of about 50,000 K, informs us directly about the temperatures of the transition region between chromosphere and corona. Located in the background, the photosphere is invisible. North at the top and East to the left. This composite image is made from 200 elementary images, obtained with the AIA instrument of the *Solar Dynamics Observatory* (SDO) mission (NASA), on November 14, 2011. (Source: Adapted by S. Koutchmy & E. Tavabi from SDO/AIA)

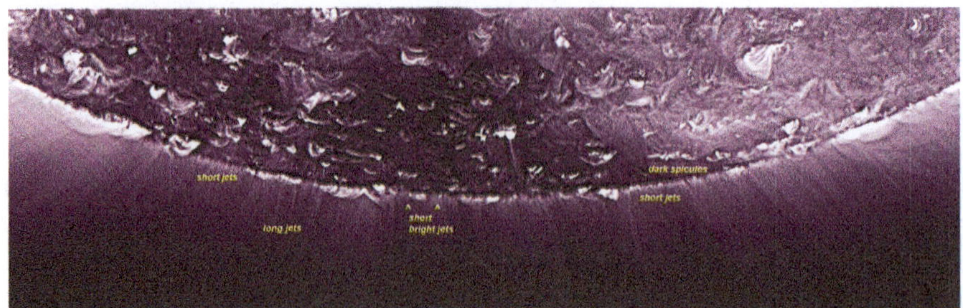

Fig. 7.21 Region of the south solar pole with coronal hole (slightly darker part on the disk) and coronal structures in emission in the 19.3-nm channel of Fe XII formed around the temperature of 1.5 MK. Image made after summing several tens of routine elementary images with the AIA instrument of the SDO mission (NASA). (Source: S. Koutchmy & E. Tavabi)

correspond to emerging magnetic fields, which release their energy and produce heating in the transition region. The large prominence, located in the Southeast (lower-left), has been the subject of extensive studies showing considerable turbulent velocity fields, suggesting forced heating causing the plasma to reach a temperature close to that of the corona. This prominence will rise far from the surface, 2 days after this observation.

On this same image (Fig. 7.20), one can see hints of a halo of coronal emission, especially at low latitudes. It is due to the emission of a coronal line, here parasitic, of ionized silicon (Si XI), less bright and emitted around two million Kelvin, at a wavelength close to 30.4 nm, transmitted by the filter and therefore contributing to this image. The intensity of this line is negligible in the structures of the transition region where the temperatures, varying from 10,000 to 500,000 K, are too low for the gas to emit this line.

On this other image (Fig. 7.21), made with a filter selecting a line of iron XII (i.e., ionized eleven times), the small cool structures, such as spicules, are seen in absorption. They are therefore dark and appear mainly at the limb of the Sun. Further from the limb and even on the disk, other small bright fine structures appear on this deep image of the corona, including Eiffel Tower-shaped jets (inverted letter "Y") that seem to feed the corona with ionized plasma by filling larger structures, known thanks to eclipse observations. These structures are then called plumes or polar jets, and it is assumed that they are confined by the Sun's main magnetic field, which exists on a large scale.

This series of monochromatic images (Fig. 7.22), taken from space in the extreme ultraviolet, once again shows the different aspects of the corona according to temperature. The image on the left is taken in visible light: it is

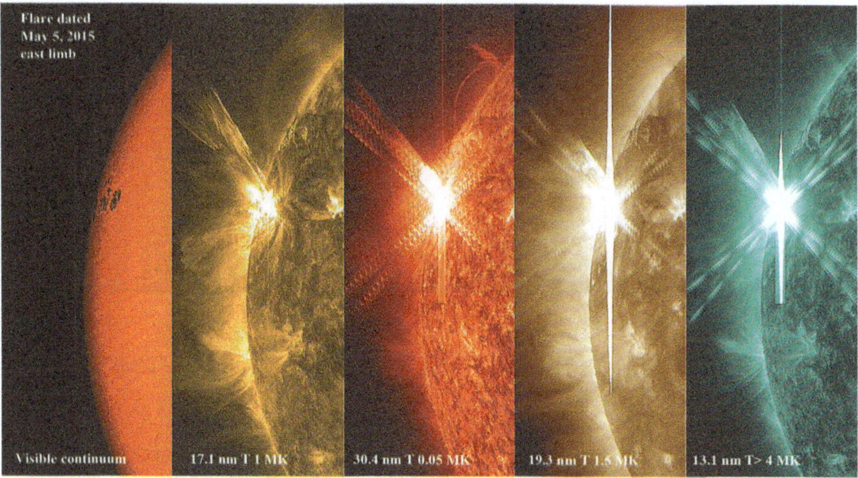

Fig. 7.22 Partial solar images with a flare on the east limb (left), in the visible and in the extreme ultraviolet. Different spectral channels are represented in arbitrary colors to illustrate the different aspects of the corona according to the temperature, increasing from left to right (except for the 30.4-nm channel), with the AIA instrument of the SDO mission Simultaneous observations on May 5, 2015. The region of the intense flare, which appears in the form of the letter X, is overexposed (saturated) in the high-temperature channels and even in the 30.4-nm channel, corresponding to light emitted by the transition region. (Source: NASA)

an image of the photosphere, on which a dark spot is clearly distinguished, which we know is home to an intense magnetic field. Above the spot, an intense flare is seen in the corona. This correspondence illustrates well the relationships existing between the different layers of the solar atmosphere, as has already been emphasized above.

Instead of only selecting a well-chosen wavelength with a specific filter to then make the image, it is also possible to build a more complex instrument, which gives a full image of the Sun for each of the emitted wavelengths. The image is then called a slitless "spectroheliogram." The example chosen here (Fig. 7.23) is made from space in the extreme ultraviolet. The spectroheliogram includes the most intense emission lines in the wavelength range between 28 and 33 nm. At the center of the image appears the line, known as the resonance line with a wavelength of 30.4 nm, due to completely ionized helium, denoted He II. The line of silicon XI is visible immediately to its left, in its wing. Other lines, from a very hot flare on the disk, are identified further to the left, including that of iron XXII.

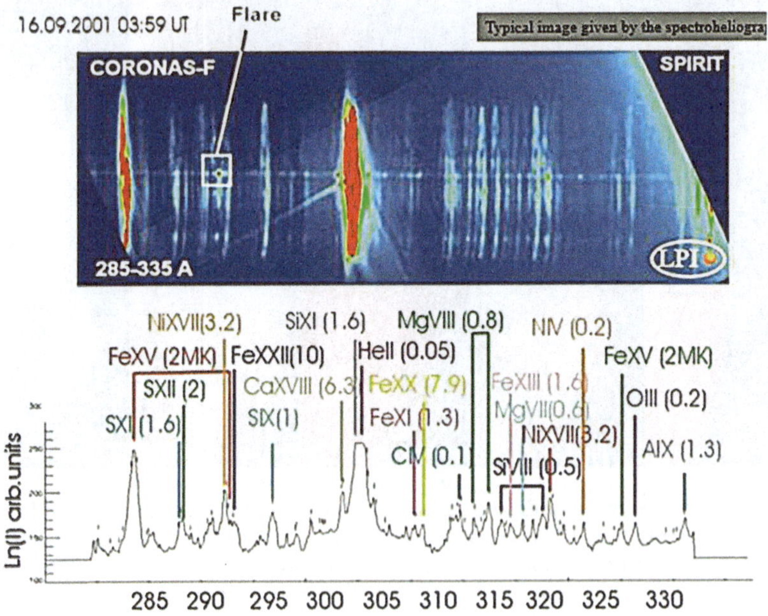

Fig. 7.23 Spectroheliogram of the Sun, made in the extreme ultraviolet and showing very elongated structures (so-called slitless spectrum). The presence of spectral lines in the image is reflected in the lower panel, which shows the relative intensities of the lines and the chemical elements that are the origin of them. These observations from September 16, 2001 were made using the SPIRIT instrument, quite unique of its kind, from the Lebedev Institute (Moscow) and installed on the satellite *Coronas-F* mission. (Source: S.K.)

Box 7.6 Why Does a Spectral Line Provide Information About the Emitting Gas?

A spectral line corresponds to the emission of light by a gas at a specific wavelength. Whether originating from the chromosphere or the corona, observed in visible light during an eclipse or outside of an eclipse by an observatory in space in the extreme ultraviolet or X-rays, this line can only be interpreted by understanding its production process by the gas atoms.

The wavelength of this line unequivocally identifies the atom (or ion) that emitted it: hydrogen or helium, abundant in the Sun, and other less abundant elements such as carbon, iron, calcium, and silicon. The relative abundances of chemical elements in the Sun have well-established reference values. However, the observation of certain lines, especially in flares, highlights enrichments that need to be interpreted. Observations made in situ in the solar wind, far from the Sun, by the STEREO mission, with its two satellites orbiting heliocentrically, crossing ahead of and behind the Earth, have confirmed these abundance anomalies.

Sometimes two elements emit wavelengths very close to each other, and separating them will be difficult, except with a high-performance spectrograph. The wavelength of the line also identifies the possible ionization state of the atom, having dissociated by losing one, two, or more of its electrons in the plasma. This dissociation provides a direct measurement of the gas temperature at the point where it emits the line. Of course, when the line is observed in a certain direction, it may result from the superposition, along the line of sight, of emissions from regions of different temperatures, making the measurement more difficult to interpret.

Once the position of the spectral line is identified, and thus the element that originated it, the measurement focuses on the intensity of the line, which directly depends on the number of atoms (or ions) that emitted it. A line is emitted when the atom (or ion), having absorbed energy, is therefore in an excited state, and then transitions to a less excited state, with the difference in energy between the two states carried away by the photon of light. Therefore, the atom (or ion) must first have absorbed energy (becoming in an excited stage). This energy comes either from absorption of light or from collisions of with electrons, these collisions themselves resulting from the thermal agitation of the gas, hence its temperature. The de-excitation of the atom (or ion), accompanied by the emission of light, is a fundamental process of atomic physics, which results in an emission probability, known and characteristic of the energy involved. The coefficients representing the probabilities of atomic transitions in a gas, by linking them to the emission or absorption of light, were introduced by Albert Einstein at the beginning of the twentieth century. In 1921, this fundamental work earned him the award of the Nobel Prize in Physics. Since then, these coefficients bear his name.

When the density of the gas is high, collisions are frequent and the spectral line directly reflects them. When the density is low, as in the high corona, collisions can become too rare to excite the ion, but another mechanism can then come into play. The light from the solar disk can provide the ion with the energy of excitation, and the ion can spontaneously de-excite by emitting the X- or UV-radiation that is observed, which is referred to as "resonance scattering."

Spectral lines also contain direct information about the agitation of the atoms or ions that emit them, when we observe and measure their profile. When the line is emitted by an atom that is in motion relative to the observer who receives it, the received wavelength is slightly modified (Doppler–Fizeau effect). The random movements in the gas will therefore broaden the spectral line, which then includes wavelengths distributed around its main wavelength: this is called the profile of the line and this distribution is a direct measure of the agitation of the gas, whether it is caused by its temperature or whether there is added turbulent agitation.

Finally, the presence of a magnetic field, which is of major importance in the corona, modifies the trajectories of the ions, which are subject to this field. While in the absence of a field, they move in a straight line between two collisions, the force due to the field superimposes a circular motion, called gyration, which may or may not modify the spectral profile, depending on the intensity of the magnetic field and the density of the plasma.

These brief comments, relating to interactions between atoms and light, show the richness of the observation of spectral lines from the chromosphere, the transition zone, or the corona, in order to better understand these regions.

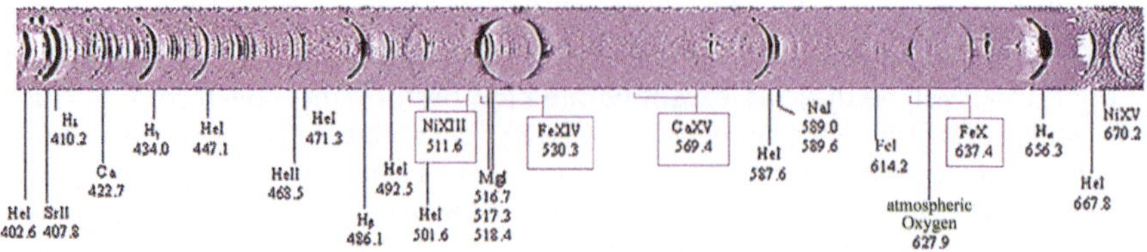

Fig. 7.24 This deep, "slitless" spectrum, of the visible spectral domain of the inner-most corona and of the transition region between the chromosphere and the corona, was taken in France by the amateur–teacher Denis Fiel (Paris), during the 1999 eclipse. The spectrograph, a light-dispersing instrument, included a grating of 600 lines/mm used in transmission, a relic of the Soviet era. The photographic spectrum, measured after computer processing, covers wavelengths ranging from blue (wavelength of 400 nm) to deep red (670 nm). The identified emission wavelengths for the corona are: the green line of the 13 times ionized iron, denoted Fe XIV (530.3 nm), showing the corona all around the solar circumference; weaker lines of nickel XIII, of calcium XV, nickel XV, and the famous red line of iron X. For the cool chromospheric lines in emission, that of Hβ (486.1 nm), of helium He I (587.6 nm), of sodium Na I (589.0 and 589.6 nm); of hydrogen Hα (red line at 656.3 nm), and of helium He I (667.8 nm). Numerous other weaker emission lines are clearly visible in the blue, corresponding to normally abundant elements that are ionized only once in the corona (This phenomenon was the subject of a PhD defended in 2014 by Cyrille Bazin, see CNRS "PhD" database, http://tel.archives-ouvertes.fr/tel-00921889v2). (Source: Denis Fiel & S.K.)

With these few examples, let us conclude that the observation of spectral lines is a powerful tool for investigating the corona, whether it is about observations from the ground, which allow access to line profiles and distant regions of the corona,[5] more than two solar radii away, or space observations in X-rays or the extreme ultraviolet EUV, more limited in precision at these distances because of the modest size of the telescopes used (Fig. 7.24). By their precisely measured position, these lines provide measurements of the gas displacement speed, while their wavelength profile measures its temperature as well as the turbulent agitation it may undergo, including when propagation or standing waves are produced.

7.5   Coronal Dynamics and Solar Wind

Although the term "solar atmosphere" is somewhat paradoxical to designate these two components, K and F, of the corona, since they extend extremely far from the Sun while being extremely tenuous, we have used it to

---

⁵ Koutchmy et al. (2019).

designate this immense cloud of plasma and solid grains. Observations, during eclipses as well as outside eclipse, with coronagraphs or thanks to X-rays and far ultraviolet light, have shown that this coronal environment is not static: large jets, prominences at the solar limb, and continuous or abrupt ejections of matter into space are all phenomena that highlight the dynamics of the corona, and particularly of the K component. The solar wind, already briefly mentioned, deserves special attention for the fundamental role it plays in the relations between the Sun and our Earth. The same is true of ejections of matter, already encountered under the name of CMEs, or intense radiation bursts, since these events also have a direct impact on all the space where not only the Earth navigates, but also on the humans sent into space as well as a multitude of satellites equipped with delicate electronic and computer systems.

A Historical Glance

The totality of an eclipse lasting only a few minutes, seven at most in rare cases, the first eclipse observers only obtained an instantaneous vision of the corona. One can imagine the discussions that ensued on the association, or not, of an aspect with the Sun. When in the nineteenth century expeditions to various sites of the totality band, which extends over thousands of kilometers on the surface of the Earth, first drew, then photographed what was seen, sequences of images of the corona, separated by several tens of minutes or more, were obtained. They first proved that the prominences, then visible on the solar limb, indeed belonged to the Sun. The names of the astronomers Angelo Secchi, Jules Janssen, and Norman Lockyer remain attached to this fundamental result, established notably during the 1859 eclipse. With the development of photography and appropriate optical filters, these multi-site observations highlighted significant speeds of movement of the observed structures, up to several hundred kilometers per second. Thus, during the eclipse of 1973 and in cooperation with the Kiev Observatory (Ukraine), Serge Koutchmy and his team observed it from Mauritania and Chad—at different times therefore.[6] Similarly, during the 1991 eclipse mentioned above, whose totality zone covered Hawai'i (Fig. 7.12) and then reached Brazil, this same team collaborated with the National Solar Observatory of the United States (NSO) and managed through these observations at two sites, which can

---

[6] Numerous publications in Russian, such as Sov. Phys. Dokl. 19(10):613 (1975). See also (Koutchmy et al. 1973).

be described as stereoscopic, to reconstruct coronal structures in three dimensions.[7]

Thus, gradually from the beginning of the twentieth century, relying on the large jets directed toward the interplanetary medium (Fig. 7.2), the image of gas flows escaping radially from the Sun was forged among eclipse observers.[8] Since the plasma corona was confined to the vicinity of the Sun by the magnetic field, how then to eject this gas (called "streamers") at supersonic speed in these jets?

There was little doubt then that the anchoring, in the solar photosphere of the "magnetic skeleton" of the corona imposed on coronal structures to follow, more or less rigidly, the rotation of the Sun. The existence and appearance of large prominences embedded in the corona suggested forces of magnetic origin. The accumulation of a large number of eclipses, observed over the years, clearly showed that the overall appearance of the corona was subject to the 11-year solar activity cycle. Even more convincing were the correlations observed between, on the one hand, the magnetic activity manifested by sunspots and faculae and, on the other hand, the disturbances of the magnetic field, observed at Earth, or the auroras. As a time-lag of 3–4 days was observed between the two phenomena, this implied, for the supposedly ejected matter, a very high propagation speed between Sun and Earth. A value of several hundred kilometers per second was required, which therefore proved to be supersonic in the coronal plasma.

For a long time, it was suggested that so-called M regions existed on the Sun to explain the source of geomagnetic disturbances. These mysterious regions "M" have been searched for decades on the Sun by examining the correlations between positions of candidate regions on specific solar longitudes, therefore attached to solar rotation and possibly disappearing for the observer, and subsequent manifestations of their effects on Earth, with endless controversies because correlations were sought with active regions and surface flares. It is proposed today to consider coronal holes as the main culprits of the effects attributed formerly to the "M" regions, adding that these are regions far from active regions and possessing a dominant magnetic polarity over their entire extent. This last configuration produces open field lines in the interplanetary medium. The case of polar regions during periods of minimum activity is particularly convincing in explaining the geomagnetic disturbances that occur on Earth in a recurrent way during the years of minimum solar activity.

---

[7] Zirker et al. (1992) and the first 3-dimensional observation of structures: (Koutchmy and Molodensky 1992).

[8] The large jets or streamers: are analyzed in (Koutchmy et al. 1974), as well as in Fig. 7.1.

The Solar Wind

The German physicist and astronomer Ludwig Biermann postulated around 1950 the existence of ionized gas flows between the Sun and Earth, as well as a mysterious "radiation" initially called corpuscular radiation. The observation of the tails of comets approaching the Sun seemed to confirm this hypothesis. When a comet approaches the Sun, its heated nucleus releases grains that, under the effect of the radiation pressure of the sunlight, are the cause of the beautiful dust tail in the direction opposite to that of the Sun. On the other hand, a second tail is observed, oriented in a different direction from the first. This angle reflects the effect of a "wind" of particles acting on the ions, emitted by the cometary nucleus and therefore forming the second tail, whose speed is added to that of the cometary nucleus. This angle being about 6 degrees (one tenth of a radian), it is deduced that the speed of the "wind" acting on the ions is ten times greater than the speed of the cometary nucleus. Moreover, the very long gaseous tails of comets, when they approached the Sun within the Earth's orbit, often showed sudden disturbances, outbursts, sinuous structures, and detachments of matter that could hardly be explained otherwise than by the influence on these tails of a magnetized fluid flowing at high speed and itself inhomogeneous in density and speed.

The solar wind was then more accurately viewed by a long list of physicists, theorists of cosmic plasmas and the corona, in Europe, the USA, and the Soviet Union.[9] As early as 1958, interplanetary probes were sent into space, measuring the speed and density of the gas in which they moved, and these calculations found their confirmation (Fig. 7.25).

The most popular of the theories was due to the physicist Eugene Parker (United States) who convincingly described this solar-wind phenomenon in 1958 and then in 1964. He established a model of a homogeneous spherical corona where the movement of the coronal gas simply obeyed the fundamental laws of physics: to conserve mass, momentum (the product of mass and speed), and energy. The still very limited measurements of near-Earth probes indicating supersonic speeds of the gas were definitely confirmed[10] by the

---

[9] Apart from L. Biermann in Germany, already mentioned, notable historical contributions were made by the British S. Chapman, V. Ferraro, T. Cowling, D. Blackwell, A. Hewish, the Europeans H. Alfvén, H. Van de Hulst, K. Kiepenheuer, J. Lemaire, A. Dauvillier, R. Michard, the Soviets I. Shklovsky, S. Pikel'ner, S. Vsekhsviatsky, G. Nikolsky, and the Japanese K. Salto, notably, mixing precise but limited quantitative contributions and rather descriptive and theoretical speculations that are rather qualitative, global, and risky.

[10] Snyder et al. (1963).

Fig. 7.25 Comet West 1975, observed on March 12, 1975 with a blue filter, to show the angle between the gaseous ionic tail, which is subject to the pressure of the solar wind, and the dust tail, subject to radiation pressure in a more radial direction. The precise value of the angle between these two tails depends on the geometry of the comet's orbit. (Source: S. K.)

mission Mariner II. They were in agreement with this model as with the previous conclusions on the terrestrial effects of matter coming from the Sun. Obviously today, we know that the plasma corona consists of structures at different scales and the hypothesis of a homogeneous spherical corona was a useful, but excessive, simplification.

Today, there is no longer any doubt that the Sun loses matter—mainly electrons, protons, and ionized helium atoms—constantly, at the rate of about a million tons per second. This matter fills the interplanetary space with an extremely thin plasma, where these particles are therefore distant from each other and do not collide with each other—and which move away from the Sun from where it originates at high speed (Fig. 7.26).

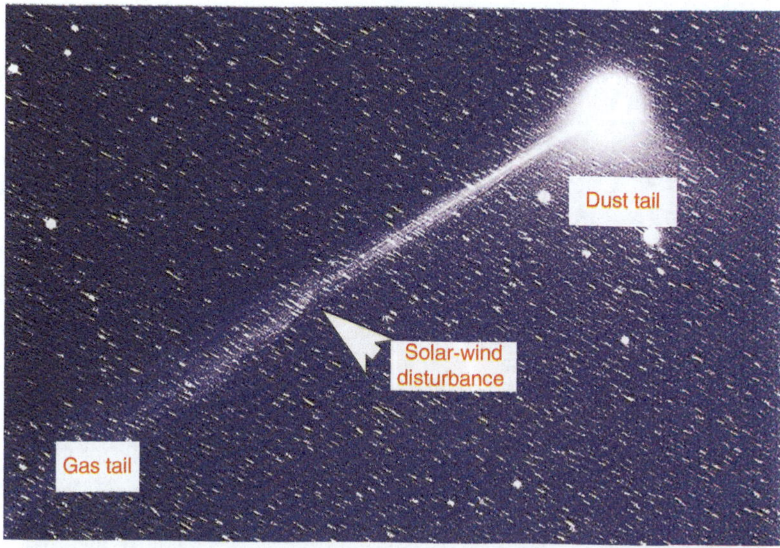

Fig. 7.26   Comet Linear 2002, observed on May 20, 2004 as it moves away from the Sun after its perihelion passage. An ionic gaseous tail, quite straight, has developed and it clearly highlights the influence of the solar wind that disrupts the flow of ions (charged molecules). The dust tail is then hardly visible. (Source: Christian Valdrich)

Wind and Clouds

In 1990, an interplanetary probe named *Ulysses* was launched by a cooperation between NASA and the European Space Agency. For the first time, this probe left the ecliptic plane and could therefore observe the Sun not from the "front" as seen from Earth, but by aiming at its North and South Poles, from the probe's polar orbit. It remained quite far from the Sun, beyond one astronomical unit, and the duration of its mission covered two solar cycles between 1990 and 2009. Its measurements thus compare the interplanetary medium, in density, temperature, and speed, during a period of low activity in 1994–1995 and then 6 years later in a period of high solar activity. The direct evidence of two types of wind, one called "fast" because it can reach 800 km/s and the other called "slow" because its speed is half or even less (Fig. 7.27), was then a surprise, because only the fast wind was expected following previous observations of effects on Earth.

In fact, with the solar telescope of the National Solar Observatory (Kitt Peak, Arizona), previous measurements of the magnetic field on the surface of the Sun had highlighted large areas of the surface where the magnetic field was uniformly directed in one direction (unipolar, North, or South). The lines

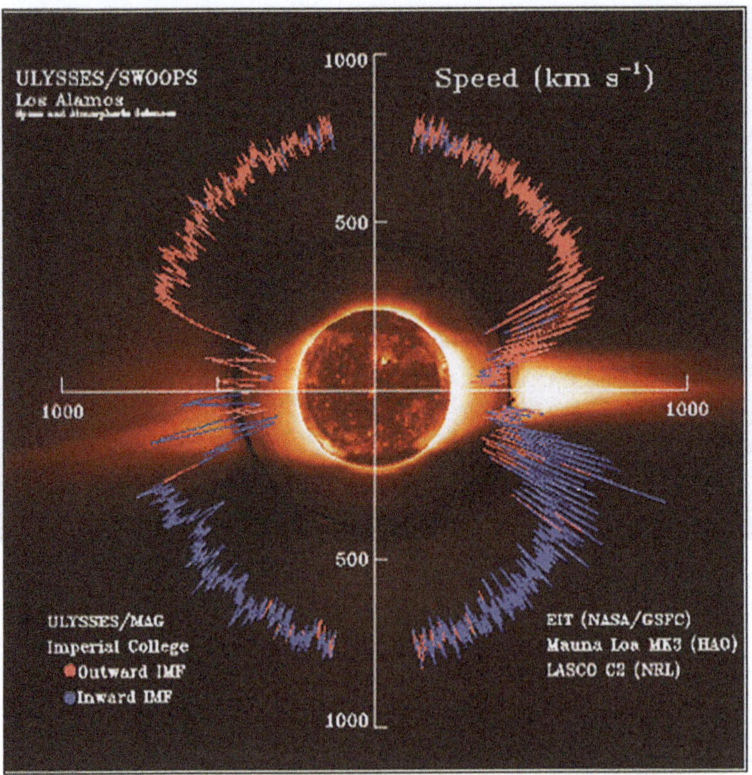

Fig. 7.27 Fast and slow winds. This figure shows the distribution, around the Sun and at a distance of about 1 AU of the measured wind speed according to latitude. In red, north latitudes. In blue, south latitudes of opposite magnetic polarity. The values of speeds in km/s are indicated on both axes. The measurements were collected in 1994–1995 during a solar minimum by the *Ulysses* probe (ESA–NASA). The central image of the Sun, also during a minimum, is on a different scale and shows the surface and the low corona (McComas et al. 2002). (Source: Courtesy of D.J. McComas)

of force of the field, anchored in these regions and open toward the interplanetary medium, allowed the acceleration of the plasma, whose particles spiraled around these lines of force, and the ejection of the plasma toward this interplanetary medium.

Note that the direction of the wind, as it moves away from the Sun, for example, in the plane of the Ecliptic, generally follows curves similar to an Archimedean spiral, because the source of the wind is anchored in a Sun rotating on itself with an average period of 27.5 days. This trajectory of the wind in space thus resembles that of a jet of water coming out of the gardener's sprinkler to reach all of the flower bed.

The Heating of the Fast Wind The role of the magnetic field in ejecting the wind into space is therefore reasonably well understood. The same cannot be said for the high temperatures reached by the plasma constituting the wind—the heating mechanism of which remains in debate. Eugene Parker had predicted that the speed of the wind was all the more significant as its temperature was high and that it could thus overcome gravity and escape the Sun's gravitational attraction. However, the coronal holes, from which the fast wind emerges, are less hot than anywhere else. There must therefore be additional sources of plasma heating. The dissipation, by turbulence, of waves propagating in the plasma in the presence of magnetic fields, remains a possibility, as well as the phenomena of sudden reconnection of the magnetic field, releasing magnetic energy and heating the plasma. Eugene Parker had thus suggested "nano-flares," which have not yet been convincingly observed but may exist.

The Slow Wind and the Micro-Structures Unlike the fast wind, which emerges from magnetic regions open to interplanetary space, the slow wind comes out of closed magnetic regions, including active regions, where the lines of force are anchored to the solar surface forming fine loops and, higher up, arches. The modeling of such regions remains difficult, even with three-dimensional numerical simulations carried out on the largest computers on the planet.[11] These regions are the site of violent and transient phenomena of great complexity and the hypothesis of a quasi-stable slow wind with spherical symmetry no longer makes much sense there.

To better understand the birth of these structures in the vicinity of the surface, a new mission of the European Space Agency, named the *Project for On-Board Autonomy-3* (PROBA-3: 2024), mainly consists of an association of two satellites flying in formation, one carrying a telescope, the other a mask located about a hundred meters away, thus producing an artificial eclipse in an external occultation coronagraph: the *Association of Spacecraft for Polarimetry and Imaging of the Corona of the Sun* (ASPIICS) experiment. For a few consecutive hours during each, very elliptical orbit around the Earth, the corona will be observed near the surface of the Sun.

To make progress today on the sources of heating of the slow wind, it is necessary to examine the detail of very small structures, where this wind has

---

[11] Very remarkable work is being done by groups like the one in San Diego created by J. Linker and Z. Mikic, see Predictive Science Inc. Coronal Modeling (predsci.com)

its source. Indeed, the tens of millions of images, collected over more than 20 years by the LASCO coronagraph of the SOHO mission, form time-lapse movies. They show the changes in the dense structures of the K corona, several solar radii from the surface and over the course of a solar rotation.[12] They leave no doubt about the existence of a myriad of small, expanding structures along the jets and at the edge of the jets.

Streams of small plasma clouds move in the wind, guided by stretched invisible force lines but still anchored mainly to the surface of the Sun from where they originate. Alas, today the inner corona remains totally hidden by the masks of space coronagraphs, up to a radial distance of 1.3 radii from the solar limb.

In the laboratory, small clouds forming in a plasma subjected to a magnetic field have already been observed and the fact that such a process could occur in the corona is not unrealistic. It led to the idea of "plasmoids." But it is the observations of the morphology and source of such clouds in the corona, as close as possible to the solar limb, that are missing, in order to understand the mechanisms at work. We have already mentioned an exceptional observation (Fig. 7.12), thanks to the use[13] of the large *Canada-France-Hawaii* telescope, which was in the band of totality of the 1991 eclipse. Other fast cameras, also installed on this telescope, have searched for the "nano-flares" predicted by Parker to heat the corona and accelerate the slow wind, without observing anything that resembled them. In 2022, the temporal and spatial resolutions of the 1991 observations remain unmatched, whether on the ground or in space (Fig. 7.28).

Since the first observations of coronal images made at wavelengths of X-rays and far ultraviolet by the NASA *SkyLab* mission, bright points, distributed over the entire surface, including in the polar coronal holes, had been observed. The numerous missions that followed, already extensively cited concerning the K corona, have improved the spatial and temporal resolution of images of the solar surface (Japan's *Yohkoh* and *Hinode* missions, NASA's TRACE and SDO missions, Europe's *Solar Orbiter* mission).

The bright points are resolved into typical magnetic structures, with small-scale explosive phenomena that produce plasmoids as well as directed jets. The existence of these processes, observed on the surface of the Sun, is hardly in

---

[12] SOHO Science Archives, https://soho.nascom.nasa.gov/data/archive/.

[13] These observations are described in (Koutchmy et al. 1993). For the observation of coronal threads of 0.4 arcseconds in diameter, see (Koutchmy et al. 1995)). These observations have been used for PhDs by Delanée C. and Zhukov A., as well as in the important work of L. November in the USA (November and Koutchmy 1996).

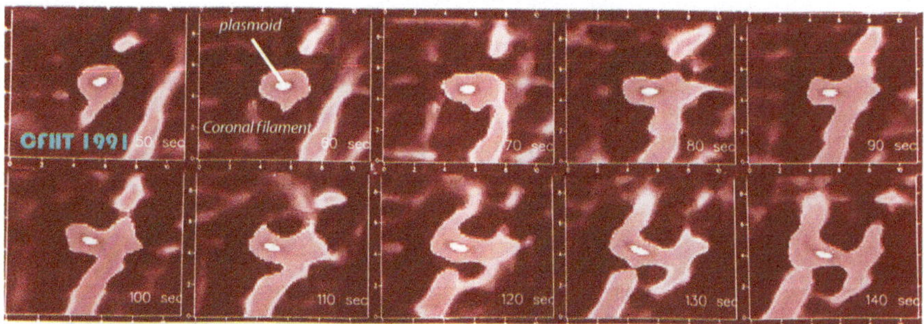

Fig. 7.28 Coronal plasmoid interacting with a very fine structure observed with the optical telescope *Canada-France-Hawaii* of 3.6 m in diameter, during the total eclipse of 1991. The light, selected by an interference filter, is that of an iron X-ray emission line. The sequence of successive images, at a rate of 30 images per second, covers a small field of 6 arcseconds on a side (or one three-hundredth of the diameter solar), which remains centered on the plasmoid. The resolution of the details is exceptional, reaching 0.4 arcsecond. (Source S. Koutchmy)

doubt anymore. However, their total number on the entire surface of the Sun remains too low to maintain the high temperature of the corona and especially to supply this entire corona with particles—electrons, protons, helium nuclei—and to compensate for the mass carried away by the solar wind that escapes in a continuous flow from the Sun. The bright spots certainly fascinate theorists[14] but they are only epiphenomena and do not seem to be the secret of coronal heating. Therefore, we must look more closely at what is happening on larger scales in the hope, perhaps illusory, of discovering new mechanisms.

Can we hope to see more clearly thanks to numerical simulations, where we try to reproduce in the calculations of a computer the complexity of the physical situations encountered in the lower corona and that are better and better known by the measurements of observations? This is the path followed by researchers from the Japanese observatories in Tokyo and Kyoto, simulating in two dimensions the birth of directed jets observed in X-rays by the *Yohkoh* mission (Fig. 7.29). They subsequently carried out three-dimensional simulations, which require more powerful computing resources. These have been able to make great progress in understanding the mechanisms at work on small scales.[15]

---

[14] Yokohama and Shibataka (1998) and the 2D-theoretical model, at high resolution, of a mini-bright coronal point developed with large numerical methods in the team of F. Moreno-Insertis at the Institute of Astrophysics of the Canaries in La Laguna-Tenerife, Spain: Nobrega-Siverio et Moreno-Insertis (2022).

[15] Yokohama and Shibataka (1998) and K. Shibata at kyoto-u.ac.jp.

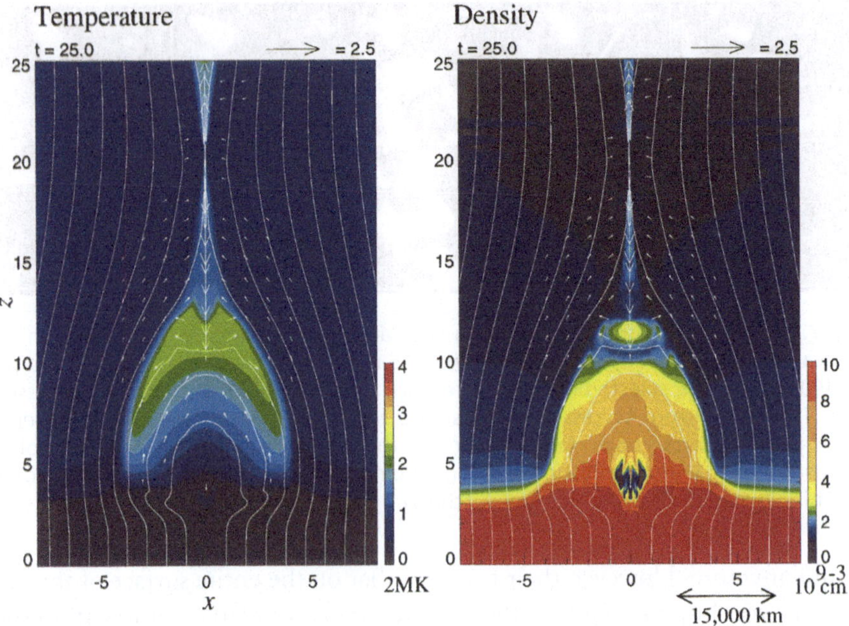

Fig. 7.29 Image extracted from a two-dimensional numerical simulation of coronal evolution with time—along the solar limb in the photosphere and perpendicularly toward space above the solar surface. Unit of distances in abscissa is 3000 km. The distances are in thousands of kilometers. The simulation calculates the temperature (left) on a scale 0–4 MK and the density (right) on a scale 0–10 particles/cm^3 at each point, taking into account the magnetic field whose lines of force in white are sometimes open to the outside, sometimes closed in loops, sometimes in reconnection with ejection of a jet of plasmoids toward the outer corona. The small arrows indicate the movements of the plasma. The elapsed time (marked t) is 7.5 min after the initial stable state (calculation carried in 25 time steps of 18 s each). (Source: Yokohama and Shibataka 1998, reproduced by permission of American Astronomical Society)

Flares, Eruptions, and Coronal Mass Ejections (CMEs)

Since the last few decades and outside of eclipses, the acquisition from space of images of the corona, on the one hand in X-ray and extreme ultraviolet (EUV) radiation with the missions *Yohkoh*, SOHO, TRACE, STEREO, and SDO already extensively mentioned, and on the other hand in visible light with externally occulted coronagraphs, such as the LASCO instruments of SOHO and SECCHI on the STEREO mission, the knowledge of the dynamics of the gas within the corona has made a colossal leap. The knowledge of the Sun, as we have recalled, owed much to spectroscopy, particularly the red line of hydrogen (Hα) emitted by the chromosphere and observable outside eclipse, as well as an intense line of calcium seen on the disk. Since the XXth century until today, the Paris Observatory in Meudon has significantly developed these studies. Later, as mentioned above, the impact of solar events on geophysics—Earth's magnetic field, radio communications, auroras—drew attention to the magnetic field on the surface of the Sun with the establishment of magnetic maps, called magnetograms, as well as on the coronal holes, observed by spectroscopy above the limb of the disk and around its entire circumference from the Pic du Midi Observatory.

The impact of the new data, resulting from space missions, is such that a new discipline, called heliophysics and already mentioned, has been born. It deals more globally with the Sun, both for itself and for all of the relationships that it maintains with its more or less close environment in the solar system, and particularly with the Earth, under the term "space weather." The multiplication of missions in near or distant space offers this heliophysics a field of development all the greater as there remain many unknowns in the functioning of the Sun and its corona.

Two Major Events on the Sun On August 21, 1996, while the Sun was at a minimum of its spots' activity, the camera observing in the far ultraviolet and the LASCO coronagraphs of the SOHO mission launched 8 months earlier, observed for the first time and for several hours a gigantic coronal eruption, with plasma mass ejection, known as a coronal mass ejection (CME), outside the ecliptic plane. We will continue to use the acronym CME to refer to these ejections.

This image (Fig. 7.30) shows the "heliospheric sheet" that separates the north and south magnetic polarities of the Sun, considered in a first approximation as a two-pole magnet (see Fig. 7.8). Beautiful vortices, which contain plasmoids, unfold at a great distance from the solar surface. These plasmoids remain visible up to a distance from the solar limb of 30 solar radii, outside the field of the image reproduced here. Even in the absence of notable solar activity on the disk, where one does not discern either spots or faculae, CMEs can therefore occur in the corona, due to large-scale instabilities. In fact, a small remnant of the prominence that is at the origin of this event is still visible at the solar limb, in the upper-right quarter.

In the early days of solar physics, eruptions in the form of prominences were already observed by Bernard Lyot and visualized in the beautiful film *Flames of the Sun* (1957), mentioned in the first part of the work about this great researcher. They occur when an image of the chromosphere is made

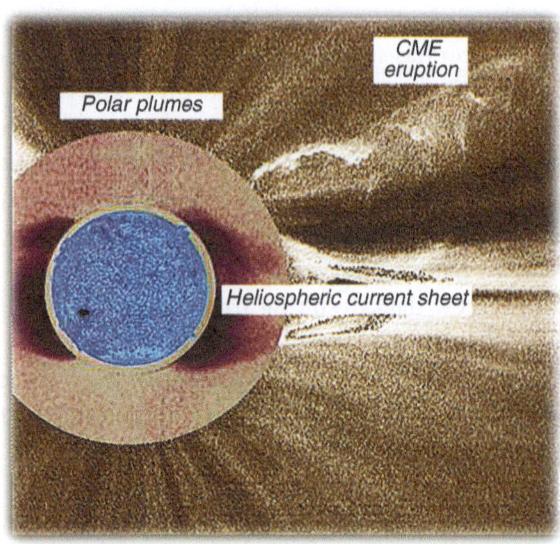

Fig. 7.30 The first coronal eruption, observed in white light by the European mission SOHO, on August 21, 1996, source of a CME observed in full solar minimum of activity. Three images are combined here: the emission in ultraviolet radiation from the disk (30.4 nm); the emission of the inner corona in the green line of the iron XIV emission, obtained by the internal occulter coronagraph, here in negative to better see the contrasts; the emission of the external corona, in positive, seen by the external occultation coronagraph. (Source: S. Koutchmy)

with a filter selecting the intense red line of hydrogen, called the Hα line. These large eruptions of tormented gaseous clouds, at a temperature still close to that of the photosphere, mix hydrogen atoms with ions and electrons. They cross the corona at high speed, dropping part of the partially ionized matter back to the surface, while the other part is lost toward the outer corona and the interplanetary medium. The phenomenon is observed with great detail from space in the extreme ultraviolet, the coolest matter being seen in absorption on the background of line emissions on the disk. The lines of helium II ions (atoms that have lost an electron) highlight on the images and in a very contrasted way, the filaments on the disk and the prominences in the corona.

Fifteen years later, on August 31, 2012, NASA's *Solar Dynamics Observatory* mission obtained an image (Fig. 7.31) where a very intense light burst, called a flare, is also manifest.

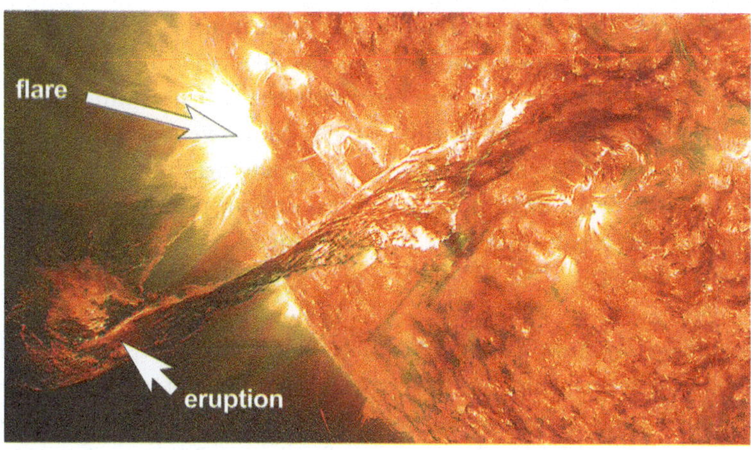

Fig. 7.31 A part of the Sun, observed on August 31, 2012 in the far ultraviolet, with a filter selecting the wavelength of 30.4 nm, emitted by the intense spectral line of the helium II ion. (Source: AIA imager of NASA's SDO mission)

On the surface of the Sun, this image shows a filament that becomes, beyond the limb of the Sun, the eruption of a huge prominence, showing vortices like the previous one. In addition, in its vicinity but on the disk, intense localized emission manifests the presence of a flare associated with a compact magnetic region. The stretching of the intertwined magnetic force lines can be suspected in the part located in front of the flare while the tearing of the structure in the upper part is apparent. Some parts, even suggesting violent turbulence, fall back on the Sun toward the flare.

The colossal radiation emitted by this flare, as by others, is also observed in numerous spectral lines emitted by a high-temperature plasma, such as that of the iron XXIV ion, which translates to a temperature exceeding ten million degrees, as well as in an intense flux of gamma rays (γ). In the case of certain flares, especially observed during periods of high solar activity, the luminous energy emitted in X-rays and the extreme ultraviolet is so considerable that, when it reaches Earth and is absorbed at high altitude by the ionosphere, it practically doubles the temperature of the latter.

It is understandable that before obtaining such images at these wavelengths from space, solar physicists, who only had very limited observations of this flare phenomenon, which is astonishing for the energy that it suddenly releases, explored all sorts of hypotheses about its cause. Great physicists such as the Soviet Armenian Viktor Ambartsoumian (1908–1996) invented particles of very high energy rising from the depths of the Sun, while the British physicist Stephen Hawking (1942–2018) proposed micro-black holes surfacing and releasing the energy of the flare. Others proposed a localized nuclear fusion, like a hydrogen-bomb. One now considers that the origin and especially the cause of the triggering of flares is to be rather sought in the corona. Moreover, there are traces on Earth of even more intense irradiations of solar origin, such as the presence, in ancient rocks or in polar ice cores, of isotopes of carbon or beryllium atoms, which can only have been produced in a distant past by such high energy fluxes.

Escaping from the Sun: Plasma Ejection CME eruptions are not only the cause of bright wavefronts, but they also produce jets of plasma that escape from the Sun in the coronal regions where the force lines are open. These jets, sometimes associated with solar energetic particles (SEPs), are frequent during activity maxima. In addition, the clouds ejected during a CME event carry with them the magnetic field that is "frozen" in it.

At the top of the image shown in Fig. 7.32, an intense CME rises above the solar North Pole and therefore has little chance of reaching the Earth's neighborhood. The bubble that detaches from the Sun contains arches that will dissipate in the interplanetary medium, as well as remnants of the ejected prominence, more or less surrounded by a cavity revealing the lines of force of

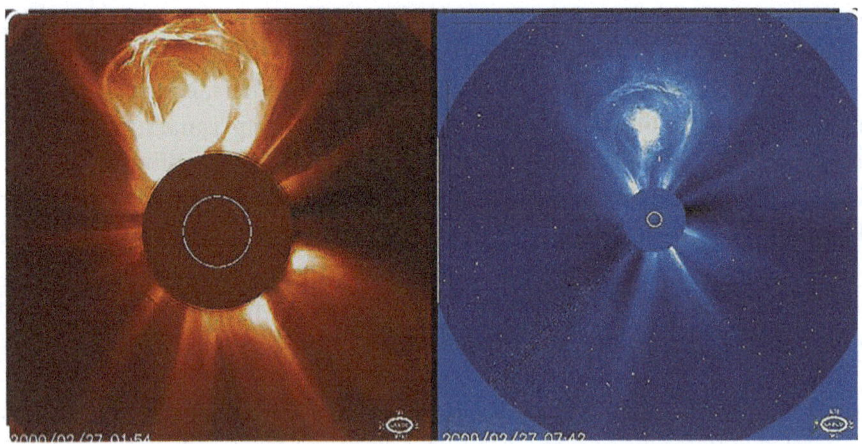

Fig. 7.32 A coronal mass ejection (CME), observed in white light, 6 h apart on February 27, 2000, by the two external occultation LASCO coronagraphs of the SOHO mission (ESA and NASA). The Sun was at that time near a maximum of activity of the sunspot cycle. The dotted white circle on the left gives the dimension of the solar disk, hidden by the mask of a radius 2.3 times that of the solar disk. In the image on the right, with a larger field of 30 solar radii, stars are clearly visible. (Source: ESA/NASA)

the coronal magnetic field. The successive images provide the speeds of movement of the CME, which vary from 500 to 1000 km/s, allowing this CME to cross the corona in a few hours. The total mass of ejected plasma is estimated at 10^{12} kg (one billion tons) or about 1% of the total mass of the corona. Here, the flare, which is at the origin of this CME, is not visible, as it is hidden by the occulting mask.

The Flare–CME Duality   Are these two spectacular events—flares and CME-independent or always connected? Is one the cause of the other? A significant flare on the surface does not always produce a CME in the corona, even of minor importance, as might be intuitively concluded.[16] However, the opposite is often observed. It appears necessary for the structure of the magnetic field surrounding the event to be open toward the interplanetary medium to allow the cloud and its procession of plasmoids to escape. Rarely, a prominence eruption occurs, associated with a more or less significant CME, without a flare in its vicinity on the surface or in the lower corona, even several hours after the event, as in the case presented in Fig. 7.30.

Understanding how the magnetic field can open toward the interplanetary medium and let plasma escape is therefore a central question. Does this result from successive flare explosions in active regions or from a slow process of evolution of heliospheric sheets? A global vision is necessary, which is not yet complete despite the numerous mechanisms proposed by specialists in recent decades. Let us try to summarize it as it stands: The ejected prominence (in red) was initially located on the surface, near the chromosphere. On it, two filaments, as visible in the red hydrogen line, are still surrounded by magnetic field lines. Stretching upward, this field is a source of magnetic reconnections, which release energy above this mass of cool matter. Heating of the plasma ejects the structure, dragging along the entire surrounding coronal region and producing a shock wave, spherical and rapidly expanding toward the interplanetary medium. Similarly, the underlying layers are heated, producing the intense radiation of the flare (Fig. 7.33).

Later, the structure contracts, producing spectacular phenomena near the surface, such as very bright thin loops and small plasmoids raining all around. These significant elements seem to be well identified in many CMEs observed for several decades, including the external spherical wave, expanding toward the interplanetary medium. It was even named the "EIT wave" after the

---

[16] Filippov and Koutchmy (2002).

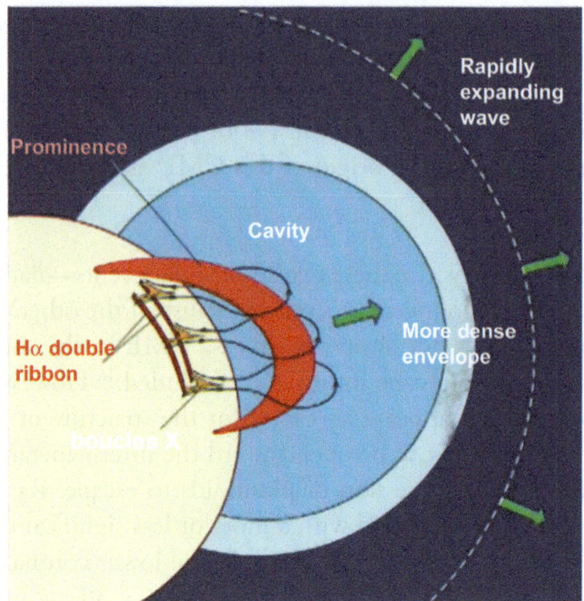

Fig. 7.33 Synthetic and speculative drawing, suggesting the mechanisms behind a CME, as proposed by many authors

Franco–Belgian instrument *EUV Imaging Telescope* (EIT) onboard the SOHO mission, as this instrument discovered this faint but spectacular phenomenon, rapidly propagating both in projection on the disk and above the limb. EIT also became the flagship instrument of the SOHO mission.

Radio-Frequency Emissions In addition to the intense emission of visible light, extreme ultraviolet, and X-rays, as well as the ejection of matter into space, we must also add the emission of this other electromagnetic radiation that is radio-frequency waves. All over the world, radio telescopes, such as those of the Paris Observatory at Nançay, observe bursts of radio-wave intensity produced by flares. The explanation probably lies in the very directed electron jets that cross the corona at speeds close to that of light and excite the oscillations of the plasma hence producing these waves. Their observation provides important information about the average corona, but is not enough to characterize the CMEs.

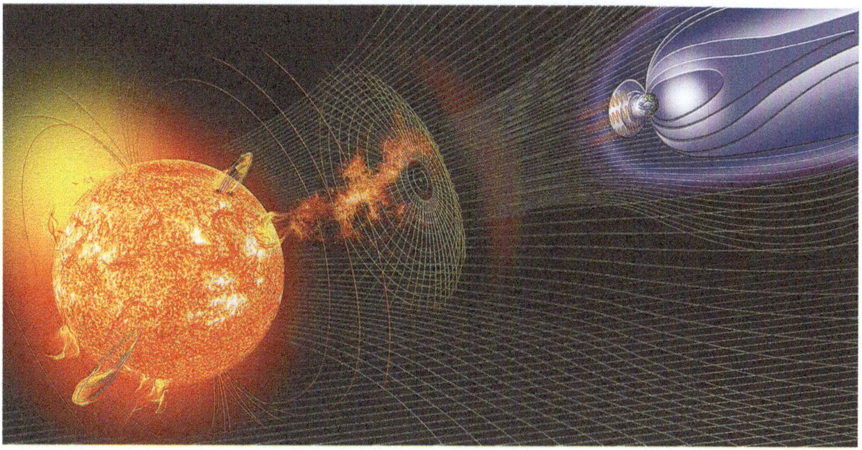

Fig. 7.34 The relationship between solar events and Earth. A plasma cloud, carrying a "frozen-in" magnetic field and associated energy, is expelled by the Sun, propagates to the outskirts of the Earth and encounters the shield associated with its own magnetic field, whose lines of force are traced (magnetospheric cavity). The plasma particles bypass the obstacle, penetrate the tail of the magnetosphere and are injected toward the polar regions, producing the beautiful polar auroras. (Source: ESA)

7.6   Conclusion

The solar corona reveals itself as a constantly changing environment, closely linked to the Sun's magnetic activity cycle, where spectacular dynamic phenomena occur, such as mass ejections and intense electromagnetic energy emissions. Matter and light propagate throughout the solar system, well beyond the Earth and the neighborhood that forms its magnetosphere. As planetary systems, different from ours, are discovered around other stars, more or less similar to our Sun, the knowledge of the solar corona and its events will undoubtedly be valuable for exploring these new worlds, their riches, and perhaps also their dangers. This is what the last chapter that concludes this book proposes (Fig. 7.34).

8

The Dust Corona and Zodiacal Cloud

In contrast to the K coronal component, made up of plasma at very high temperatures, which we have just described, the coronal component called F, which we explore here, is made up of tiny dust grains, mixed with plasma, with temperatures ranging between 1000° and 100° Kelvin depending on their distance from the Sun. The zodiacal cloud extends to the Earth's orbit and beyond. We have already mentioned the F corona in the first part of this work, as its observation near the Sun was the subject of one of the programs on board *Concorde* 001 in 1973. Observing the F corona near the Sun is difficult because, for a terrestrial observer, the brightness of the K corona dominates. Also, at that time, many interpretations of the nature, size, and movements of the dust grains were hasty or questioned. Renowned astronomers, such as the British H. Bondi, F. Hoyle, and R.A. Lyttleton in the 1950s, observing spectroscopically the excess of certain chemical elements in the corona, suggested that the grains came from far from the Sun and perhaps even contributed to heating the corona. This theory of heating by grain accretion was quickly refuted, especially when instruments placed in space allowed the determination of the dynamic properties of the K corona, presented in the previous chapter.

8.1 The F Corona and the Zodiacal Cloud

It was only from the 2010s to 2020s that space missions, approaching the Sun, brought this new perspective that we share here with our readers. In addition, the search for objects similar to Earth (called exoEarths) around

                                                                      161
P. Léna, S. Koutchmy, *Eclipsed Suns, the Solar Corona and Exoplanets*, Astronomers' Universe,
https://doi.org/10.1007/978-3-031-92199-5_8

nearby stars gives a particular relevance to these dust particles and their behavior around a star, since they play an essential role in the formation of an Earth-like planet: hence, it is the subject of the final chapter.

The light radiated by the F corona near the Sun, and then that emitted from the zodiacal cloud at a greater distance, provides valuable information about the nature of the grains, their movement, their origin, and their fate, in both of these regions. For a long time, the observation of comets has revealed the evaporation of their nucleus as they approach the Sun and the abundant production of dust that results, but we must share here the wealth of information gained in the understanding of grains, particularly thanks to space observations, since the 2000s.

Presentation of the Dust Corona

Let us examine here how this F component is known and better understood, with its dust grains.

The Faint Brightness of the F Corona

During an eclipse and just above the edge of the Sun, the brightness of the K corona dominates, then as we move away the F component takes over. However, from the first spectroscopic observations of a total eclipse, by Jules Janssen in 1869, one notes the presence of dark spectral lines, present in the white light of the disk outside eclipse and called Fraunhofer lines, as discovered in the photospheric light as early as 1820 by this physicist. If these lines are present in the light received from the near corona during an eclipse, it is therefore that a coronal component, illuminated by the photosphere, scatters this radiation toward the Earth. Subsequently these observations have been solidly supported, with the use of photography, and the F corona could be separated from the continuous background of the K corona and the coronal spectral lines, as remarkably done by C. W. Allen during the 1940 eclipse on a heavily exposed spectrum.[1] Moreover, more precise measurements could be made on spectra measured during the eclipse of May 1965, on board a NASA CV-990 aircraft. They show definitively the signature of the F corona, seen then almost "face on," up to the edge of the Moon.[2]

---

[1] Allen (1946).
[2] Koutchmy and Magnant (1973).

It is now well established that, near the edge of the Sun, the brightness of the F corona seen almost "face on" reaches about 0.1 millionth (10^{-7}) of that of the solar disk. This brightness in white light is therefore significant, it is one-tenth of that of the total corona, i.e., K + F, observed near the edge of the Sun, for example in coronal holes and outside of the spectacular jets and other loops. Modern light detectors, much more sensitive and precise than photography, facilitate this measurement of the F corona, seen from the Earth near the Sun's disk.

On this spectral cut (Fig. 8.1), just above the edge of the Moon, the characteristic lines of the scattered photospheric light appear faintly, scattered by the F corona, such as the two D lines of sodium and a line of iron. This one almost coincides with a spectral line produced by the light passing through the Earth's atmosphere—a parasitic line called a "telluric line." The lunar background obviously shows no trace of these lines. In the upper part, the main forbidden emission lines of the K corona are found, produced around 1.5 million Kelvin, with a dominant green line of iron XIV and a less marked line of iron X in the red. These spectral lines, of which the wavelengths are indicated, are called "forbidden," as they are only emitted by very dilute gaz.

Moving away from the Sun, the separation between K and F components is facilitated by the polarization of the K corona light on the one hand, and by the unpolarized light of the F corona, which largely dominates, on the other hand. As previously indicated, the K corona light results from the scattering of photospheric radiation by the coronal plasma electrons, whose motions "blur" the Fraunhofer spectrum. This separation, from observations, is

Fig. 8.1 The corona, observed just above the disk and in visible light. During the 1973 eclipse, Serge Koutchmy and Götz Stellmacher, from Moussoro (Chad), photographed the corona with a spectrograph whose slit was perpendicular to the edge of the Moon, thus covering up to 0.7 solar radius above the edge of the Moon, in the middle of the totality of the "eclipse of the century," which lasted 4 min on the ground. Figure 1.1 (Part I) is a global photograph of this corona at the same moment (Source: S. K.)

relatively easy out to a distance of 4–5 solar radii from the edge of the Sun (Fig. 8.2).

It is more difficult beyond, as the K polarization becomes weak. On the other hand, the structures of the K corona (large jets) are well marked and vary over time; here again, digital image processing allows them to be distinguished. Thus, the F component can be isolated, by making some reasonable assumptions about its light, such as its lack of polarization, its continuity with the emission of the more distant zodiacal cloud, its concentration near the plane of the Ecliptic, and the variation over time of its brightness, which is affected by the solar cycle. Also note that during a minimum of the solar cycle, the magnetic field in the K corona resembles that of a magnet buried inside the Sun, which gives a concentration of the K coronal light in the plane of the ecliptic.

In white light and moving further away from the edge of the Sun, the brightness of the F corona will gradually dominate. We have millions of images of the corona and measurements of its polarization with external occultation coronagraphs, for example with the LASCO instrument of the

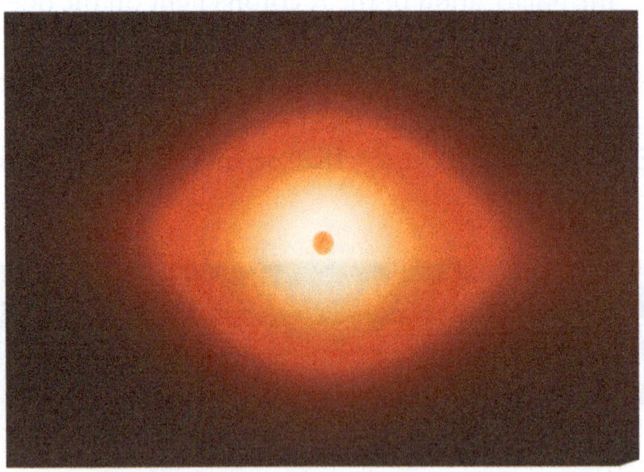

Fig. 8.2 The total corona (K + F) observed during the 2017 eclipse in the USA by N. Lefaudeux. The image is not processed, but presented in false color to better highlight the intensity variations, by compressing the digital data. In the center, the lunar disk is not totally black, mainly because of a scattered light that superimposes on it. In the outer parts, the more homogeneous F corona dominates and shows a typical ellipticity like that of the zodiacal light much further from the Sun. This presentation of the corona is rare because of the dominance of the F corona, which masks the changing structures of the K corona. (Source: courtesy of N. Lefaudeux)

SOHO space mission[3] for over 22 years, or with the coronagraphs of the STEREO probes. These images confirm the importance of the F component.

This image (Fig. 8.3) is taken from space with an external occulting mask, giving a black disk hiding the solar disk. It masks the bright parts of the inner K corona, which are sources of stray light around the arm supporting the occulting disk. The digital processing of the image enhances the contrast beyond ten solar radii and highlights the F (dust) corona, which intensely scatters the white light of the Sun. Intensity enhancements are visible on each side on the Sun's Equator and the flattening along the plane of the Ecliptic is

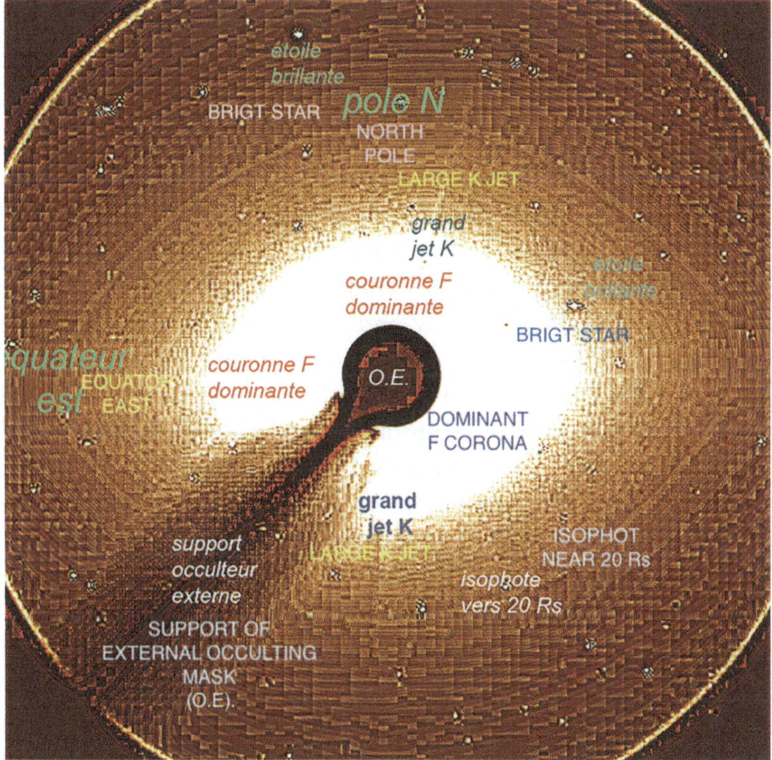

Fig. 8.3 The F corona on July 14, 2000, near a peak of solar activity. Recorded from space, in white light, by the LASCO coronagraph of the SOHO mission with its external occulting disk, this image, which extends out to 32 solar radii, precedes by one hour the triggering of a major eruption with a flare producing energetic particles in abundance for over 24 h. The event is named the "Bastille Day CME". Note: as labels in French coud not be removed, their English translation is appended to them. (Source: ESA & Image processing by S. K.)

---

[3] Lamy and Gilardy (2022).

clearly visible on the isophote lines, that is, of equal light intensity. Numerous stars are visible in the field, which clearly indicates that from space, the background of the sky is really black! The SOHO mission producing continuously several images per hour; there is a huge wealth of information, which allows extremely refined digital processing on the images.

Finally, as discussed in the first part of this work, the observation of the thermal radiation of the grains in the infrared allows the observation of the F corona further from the disk, up to several solar radii from the edge. Does there exist in this region a "grain-free zone" that is too close to the solar disk; a zone that the observation related in Part I of this work, was looking for? This hypothesis, long formulated at the time of the 1973 eclipse, needs to be closely examined, in relation to the passage of comets, now observed even very close to the solar surface. We will therefore return to this later.

Troublesome Stray Light

Distinguishing the F component from the K component is therefore possible (Fig. 8.4). However, to these two is often added stray light. Thus, during an eclipse observed from the ground, the Sun illuminates the Earth's atmosphere outside the zone of totality (Figs. 1.2 and 5.4, Part I), this light is scattered toward the zone of totality, then is scattered a second time by the Earth's atmosphere into the telescope to overlap with the image of the corona. This double scattering by terrestrial air produces a polarized stray light, which is difficult to eliminate and has sometimes led to confusion in the interpretation of the spectra obtained during eclipses.[4] By climbing into the stratosphere, as the *Concorde* aircraft did in 1973, or other flights mentioned in Part I, this parasitic light source is eliminated, but the presence of portholes in the aircraft introduces another. That is why the observation of D.E. Blackwell[5] made in 1954 from an airplane but without a porthole, during the great eclipse of minimum solar activity, has never been surpassed in quality. The raw results obtained have remained a reference even if some interpretations given by the author have been questioned. With a choice of an observation site at high altitude and by photographing in the very near infrared, the stray light due to the sky background is attenuated, as in the image of the 1970 eclipse where

---

[4] For the demonstration of chromospheric parasitic lines on a coronal spectrum during the totality of an eclipse (Stellmacher and Koutchmy 1974).

[5] Blackwell (1955).

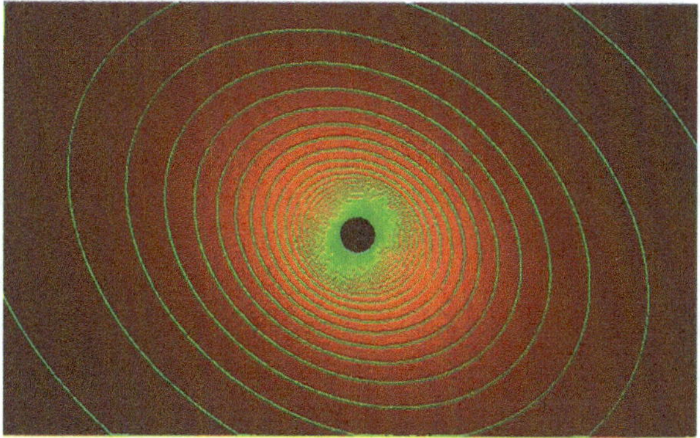

Fig. 8.4 Representation of the light intensity of the F corona in white light, moving away from the Sun. This intensity is deduced from the analysis of images taken during numerous eclipses, such as that of Fig. 7.2, by integrating them into a digital model. The solar disk, in black, gives the scale. The green lines are lines of equal intensity (isophotes), calculated by assuming them to have an elliptical shape, the axis of the ellipses being inclined, as is often seen, to the ecliptic plane in an eclipse sky. This plane is close to the equatorial plane of the Sun perpendicular to its axis of rotation. The intensities vary between 0.1 millionth of the brightness of the disk outside eclipse, very close to the Sun, and 10^{-5} millionths for the farthest part. (Source: Calculation and drawing by F. Sèvre, IAP/CNRS)

"equatorial enhancements" appear, which can be interpreted as an F corona far from the Sun, perhaps linked here to solar activity.[6]

Thus, in space and outside of an eclipse, the outer disk of the coronagraph never perfectly hides the solar disk, and the image of the corona is affected. Only sophisticated digital treatments, now of formidable precision, allow the elimination of these parasitic effects.

We can conclude this brief presentation of the F corona, as known today, by emphasizing that the origin of the grains that compose it remains to be determined.

8.2 A Pyramid in the Sky: The Zodiacal Cloud

Observed from Earth at low latitudes, for example in Chile as shown in Fig. 1.5 of Part I, the zodiacal cloud appears, after twilight or before dawn, as a kind of pointed pyramid shooting up from the horizon toward the zenith. It has thus been named the "pyramid of the sky."

---

[6] Koutchmy (1972).

Images of the Zodiacal Cloud

Photographed, its image remains affected by the Earth's atmosphere that the light has passed through. In particular, the layers of the ionosphere, around 100 km in altitude, are often faintly luminous and their light must be separated using suitable filters. Thus, from observatories located at low latitude, for example at the Pic de Tenerife (Spain) or in Hawaii, beautiful images are obtained, mainly at an angular distance from the Sun (angle called solar elongation) greater than 40°. For smaller angles, observations from space are preferable.

The interplanetary missions *Helios 1* and *Helios 2*, launched in 1975–1976 jointly by NASA and the German Space Agency DLR, approached the Sun and their photometers measured the brightness of the zodiacal cloud, taking advantage of the satellite's rotation to sweep the cloud at varying solar elongations. To avoid being dazzled by a direct view of the Sun, the sweep was done at more than 10° above the plane of the solar Equator. The measurements collected were numerous and their precision excellent, which allowed a thorough analysis of the radial distribution of dust, in particular beyond about 20° of solar elongation.

The cloud does not appear to evolve over time and seems symmetrical with respect to the plane of the Ecliptic or perhaps inclined by a few degrees with respect to it. At 60° of elongation away from the Sun, the grains contributing mainly to the zodiacal light seen from Earth are at the same distance from the Sun as the Earth is.

As early as 1982, very interesting images, in color, were obtained in space on board the *Salyut 7* station (first joint manned flight between CNES and Soviet *Intercosmos*) by the Institute of Astrophysics of Paris, in cooperation with a Soviet team.[7] These images already allowed the visualization of the zodiacal cloud by showing color nuances and therefore a variation of the properties of the grains, especially toward the red, suggesting larger grains as one moves away from the Sun. Indeed, for small grains, the color of the scattered light depends little on the size. Despite the slow rotation of the *Salyut* 7 spacecraft on itself, the sufficiently short exposure time allows the measurement of the axis of symmetry of the cloud at solar elongations between 23° and 35°. This axis of symmetry is then close to the orbital plane of the planet Venus, which possibly indicates a gravitational influence of this planet on the grains orbiting the Sun.[8] The brightness of the Zodiacal cloud is measured very precisely by a calibration of images of well-known stars (Figs. 8.5 and 8.6).

---

[7] Koutchmy and Nikolsky (1983).
[8] Nikolski (1985).

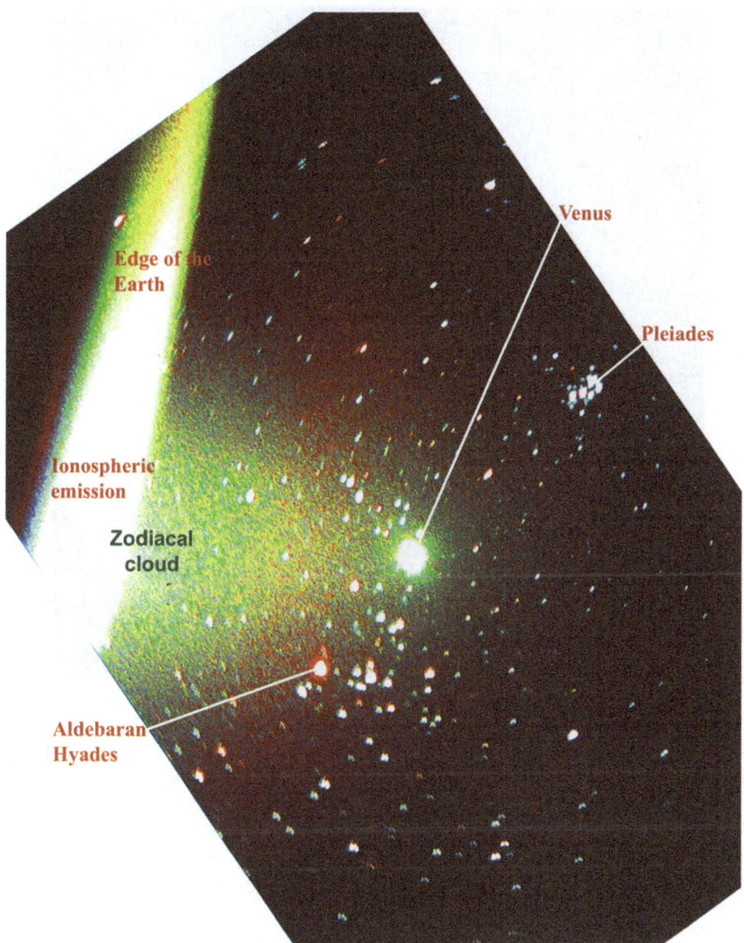

Fig. 8.5 One of the best color optical images ever obtained of the inner zodiacal cloud, outside the Earth's atmosphere, on June 30, 1982 over a field of about 45°. The PCN (Night Sky Photography) camera was installed on board the *Salyut* 7 space station. The reproduction here accentuates the contrast, which reveals the slight reddening of the zodiacal cloud, especially outside the plane of the Ecliptic, which is oriented horizontally here. The exposure and the guidance during the two-minute exposure are manually performed by aiming at the planet Venus, by the cosmonaut V.A. Djanibekov. On the left, the bright and wide trail is due to terrestrial ionospheric emissions, occurring between 80 and 110 km altitude (thus below the spacecraft), with numerous characteristic spectral emission lines identified by their color. Two-minute exposure on Ektachrome 400 ASA color film; 56 mm NIKKOR F/1.2 lens. (Source: S.K.)

Fig. 8.6 Original, unretouched, and untreated image of the zodiacal cloud, obtained from one of the interplanetary probes of the STEREO mission (NASA), far from Earth. The image covers several tens of degrees. The image of the planet Venus is overexposed, which saturates the matrix detector and produces a vertical parasitic line over the entire field. Further to the left, the Pleiades and Hyades star clusters are identified as in the previous color figure. To the right, the bright curved line results from the residual diffraction of sunlight by the masks. (Source: courtesy of C. De Forest, NASA)

After their launch in 2006 and for more than 15 years, the two STEREO probes from NASA provide a very large number of images, covering several tens of degrees. The small photographic lens of the HI (*Heliospheric Imager*) instrument, named SECCHI, is shielded from the direct sunlight by successive screens that very carefully obscure the Sun and its near corona, thanks to a device designed and built at the Liège Space Center (Belgium).

In addition to these measurements made in the visible domain, the cool grains of the zodiacal cloud reveal themselves as a source of light, this time by their thermal emission at far-infrared wavelengths (between 20 and 500 μm wavelength). This light, although faint, proves to be detrimental to essential astronomical measurements in the same spectral ranges, such as the study of the cosmic background due to the primordial universe.

The Grains of the Zodiacal Cloud

A dust grain located in interplanetary space is heated by sunlight that it absorbs and cools by radiating toward space in the infrared. It thus reaches an equilibrium temperature that depends on its distance from the Sun, but also on its size and composition. NASA's *InfraRed Astronomy Satellite* (IRAS), launched in 1982 and whose telescope had provided a complete map of the infrared sky at a wavelength of 10 microns, which was then practically unknown, had clearly detected the zodiacal cloud. Its measurements had allowed the estimation of the size of the grains to be a few microns, and led to an estimate of the number of grains per cubic meter which decreases as one moves away from the Sun, proportionally to the distance to it.

Grain, dust, asteroid: what terminology to use for these objects that travel through interplanetary space and sometimes approach the Earth's atmosphere? The International Astronomical Union (IAU), responsible for these definitions that allow scientists to agree, has officially defined a "meteoroid" as being *a solid object moving in interplanetary space, much smaller in size than an asteroid and considerably larger than an atom.* The dusty grains of the F corona are therefore meteoroids. The meteoroids that approach the Earth produce flashes and luminous phenomena by exploding or sublimating in the atmosphere can sediment in the stratospheric layers, or, for the larger ones, then called meteorites and of various masses, be collected on continents or in the oceans, thus lending themselves to laboratory analysis. A considerable amount of information is therefore available on the nature of meteoroids, indicating a very distant origin in space and time. The abundance of the chemical elements that they are made of appears to be related to the formation of the solar system and even to the evolution of the universe.

The very origin of interplanetary grains was first sought in the asteroid belt located beyond the orbit of the planet Mars where collisions between these small rocky bodies orbiting the Sun are more frequent and first produce meteoroids as debris, which would be solid bodies less than 10 m in diameter. These blocks can still break up by colliding with each other, or under the action of tidal effects, or even because of too rapid rotation if the internal cohesion forces of a friable material are insufficient.

The Movement of the Grains

Two main forces control the movement of the grains, which are in orbit around the Sun. The first is the force due to radiation pressure, which results from the intense solar illumination. It has the effect of pushing the grain outward. On the other hand, the force of gravity attracts it toward the Sun. Let us denote by the Greek letter β (beta) the value of the ratio of these forces: very small for large grains, this ratio becomes close to unity for grains with a diameter less than 1 micron, because then the pressure of light dominates. This leads to defining a family of very small grains, called β-meteoroids.

A third, more subtle force, affects the movement of the smallest grains. This is a minuscule slowdown, the appreciable effect of which only manifests after millions of years. This is the effect called Poynting–Robertson. For a rapidly moving grain, the light no longer appears to come exactly from the direction of the Sun (a phenomenon called aberration of light), which leads to a small additional force, also due to the pressure of light, but here opposed to the speed of the grain and thus slowing it down. Its trajectory is slowed down and it therefore spirals slowly as it approaches the Sun.[9]

The effect is all the more significant, at a given distance from the Sun, the smaller is the grain. Indeed, the radiation pressure exerted by solar radiation is proportional to its surface presented to the Sun and therefore to the square of its dimension, while the gravitational force exerted by the Sun is proportional to its volume and therefore to its dimension cubed. The secular evolution of the grain's orbit, due to this Poynting–Robertson effect, is very slow—millions of years for substantial debris, but the effect grows as the grain approaches the Sun and, even more, as it diminishes by evaporation when its temperature increases. For a ten-micron grain, it takes tens of thousands of years for it to significantly approach the Sun.

Two space missions highlight new and quite unexpected aspects with the new optical observations in situ from the *Parker Solar Probe* (PSP), from NASA, launched in 2018, and from the European *Solar Orbiter* (SO), launched in 2020. The PSP probe approaches the Sun, around which it orbits, as no artificial object has done before, at a distance of 5.9 million kilometers (4% of the Sun–Earth distance), protecting itself from intense radiation with a heat shield. The SO probe approaches the Sun less (31% of the Sun–Earth distance), nevertheless closer to it than the planet Mercury at its perihelion and especially the orbit of SO will slowly shift toward high latitudes to allow examination of the Sun's polar regions.

The grains of β-meteoroids, approaching the Sun, are electrically charged, either by the photoelectric effect due to the coronal X-ray and extreme ultraviolet radiation to which they are exposed or by collisions with the electrons and ions of the solar wind, which ionize them. For a long time, electric

---

[9] https://en.wikipedia.org/wiki/Poynting-Robertson_effect.

discharges, caused by the impact of these charged grains, had been observed on the antennas of space probes. But this moving electric charge has another consequence: moving in a magnetic field carried by the solar wind, the electrically charged grains undergo a new force, called Ampère–Laplace–Lorentz (the one that acts in an electric motor), which disperses them by expelling them from the solar vicinity. Finally, a fourth action can also intervene, due to their collisions with the ions of the solar wind, creating a new pressure and a new slowing effect similar to the Poynting–Robertson effect mentioned above, although more intense on the induced evolution of the orbit, because the ratio of the grain's speed to that of the wind is much higher than the ratio of the grain's speed to that of light.

These four forces, the relative importance and final effect of which depend on the size of the grain and its distance from the Sun, determine the composition and evolution of the zodiacal cloud, once the different sources of the grains can be specified. Next to the classic source, which describes these grains as resulting from collisions within the asteroid belt, a second source, directly linked to comets approaching the Sun, is beginning to be better described and understood.

8.3   Visiting and Often Suicidal Comets

We will not be able to address here the richness of the study of comets, which constitutes an entire chapter of modern astrophysics. These objects, by their intrusion into the zodiacal cloud, play an important role in the grain balance, notably by their interaction with the Sun, which they approach, and by their often immense and spectacular comas or tails.

It is the numerous images from the HI instrument of SECCHI, on board the STEREO mission, which have allowed the discovery of surprising activity in the interplanetary medium when it is crossed by comets, which are a priori foreign to the zodiacal cloud. These imagers have swept regions all around the Sun. Aiming obviously differently from that of an observer on Earth, they have collected images of comets crossing the zodiacal cloud, including Sun-grazing comet debris, members of the Kreutz family, whose orbit then brings them toward the F corona.

This impressive collection of images clearly confirms the release into space, by comets, of large quantities of grains that then feed the zodiacal cloud. Thus, shortly after the launch of the STEREO probes at the end of 2006, they were able to observe the arrival of the great Comet McNaught, which does not belong to the category of Sun-grazing comets. It was seen by one of the coronagraphs of the SOHO mission, present in space since 1995 (Fig. 8.7).

Observed a little later from the site of the European Southern Observatory (ESO) in the Atacama Desert in Chile, this same comet offered the magnificent spectacle of an immense dusty tail filling the sky and supplying dust to the zodiacal cloud (Fig. 8.8).

This comet was classified as new, having never been observed before. It therefore comes from a very distant region of the Sun, considered a reservoir of comets since the formation of the system, and called the Oort cloud.

This Oort cloud is supposed to be a large belt containing a myriad of solid bodies that are in distant orbit around the Sun. These billions of objects of all sizes can become cometary nuclei by intrusion toward the outer planets of the solar system, especially toward the planet Neptune which can deflect them

Fig. 8.7 One of the first images obtained with the SECCHI imager of the STEREO-B probe, on January 11, 2007, showing the curved dusty tail of Comet McNaught in a part of the field including the zodiacal cloud. The streaks in this dusty tail are the signature of grain sources coming from active elements of the slowly rotating nucleus. The second cometary tail, thinner, gaseous, and straight and of lower intensity, can be barely seen nearby. The comet was located at a distance from the Sun of about 40 solar radii, near its perihelion. The nucleus of the comet, and the Sun out of the image, are on the right, hidden by the mask with a vertical edge. (Source: NASA)

Fig. 8.8 Wide-field, color image of Comet McNaught, seen on January 20, 2007, in the southern hemisphere sky, during twilight. (Source: ESO)

from their trajectory. This still hypothetical cloud would be located at the confines of the solar system, at a distance from the Sun of 30,000 astronomical units, or thirty thousand times the average Earth–Sun distance. This cloud is a remnant of the formation of the solar system and its central star, the Sun, some 5 billion years ago.[10]

Originating from this cloud and after a journey of several million years approaching the Sun, the volatile elements of a cometary nucleus, such as water or carbon-dioxide ices that are present in solid form, evaporate. From the surface of the nucleus, they then release tons of grains, which form the cometary tail (Fig. 8.9). Its structures are produced by the rotation of the nucleus itself, which exposes it to a varying solar illumination and is more or less active in the emission of grains. This results in a fascinating cosmo-chemistry, which it is not possible to address here.

These examples, and many others, prove the cometary origin of at least part of the grains of the zodiacal cloud and perhaps even more of the grains of the F corona at a small distance from the Sun. In light of this observation, it becomes very interesting to discuss the observations of comets grazing the Sun and their consequences on the F corona.

It should be noted that nothing prevents a comet from entering the so-called forbidden or free zone around the Sun, which was long defined as

---

[10] https://en.wikipedia.org/wiki/Oort_cloud.

Fig. 8.9 Double image of Comet Bradfield 2004, seen during its perihelion passage by the C3 coronagraph of the LASCO instrument, on board the SOHO mission (ESA and NASA). This is located in the solar corona, at a distance of 36 solar radii from the edge of the Sun. The image is processed to subtract the F component without structures; the jets of the K corona are visible on a field of more than 30 solar radii with intensities much lower than that of the dusty tail of the comet. After this passage, this long-period comet, which does not belong to the family of grazing comets, was well observed from the ground, showing a very extended, dusty tail in the interplanetary medium. (Source: ESO/NASA)

starting from 4 solar radii and whose importance turns out to have been considerably exaggerated, if not imaginary. Comets can even graze the solar surface and then continue their journey without changing their orbit!

In 1965, the great Comet Ikeya–Seki demonstrated this by its intrusion close to the Sun, grazing its surface at 0.7 solar radii. Spectra of its coma (designation of the bright cloud immediately close to the cometary nucleus) were obtained as it passed through the inner solar corona. Showing in particular in saturated emission the double spectral line, of yellow–orange color, due to the sodium element (Na I line) and very intense in emission, these spectra confirmed the important effects of solar radiation on the nucleus, such as surface heating and evaporation, associated with tidal effects on the rotating nucleus. Subjected to these, the comet had indeed broken into several pieces, while continuing to be observed after its perihelion passage. Other observations of comets, made from the ground including during total eclipses, have noted similar intrusion phenomena.

A new era opened in 1979, with the space observations of the first small external occultation coronagraphs on the P-78-1 (SOLWIND) mission, then on the NASA SOLMAX mission, and finally in 1973 on the large *Apollo Telescope Mount* (ATM) solar laboratory called *SkyLab* and placed in Earth orbit.

In 1982, a US Navy observation satellite, equipped with a Lyot coronagraph with external occultation, detected the explosion of a cometary nucleus plunging to less than one radius from the edge of the Sun. Although the coronagraph's mask covered this region, the cloud formed by the dust grains thus released extended up to ten solar radii and became observable in the coronagraph's field for more than 20 hours, beyond the mask's limit and outshining the structures of the corona.[11] At the time, this unexpected observation was surprising. (Anecdotally, this satellite was probably the first to be destroyed by the militarization of space, as it was shot down later by a US Air Force missile.) (Fig. 8.10).

After 1996, the coronagraphs of the European SOHO mission (in close cooperation with NASA) brought a plentiful harvest of images, now obtained continuously over the years. The role of cometary activity for the F corona can therefore be analyzed with many image sequences. The K and F components are separated by image processing and the stray light of instrumental origin, considered slowly evolving over time, is eliminated.

These digital treatments allow substantial amplification of the contrast and identification of everything that changes quickly from one image to another. Thus, many objects that pass through the field of the coronagraph are discovered, including small, so-called, grazing comets, which approach the Sun to less than a solar radius from the surface and many then disappear into the corona. These comets have a nucleus of a few meters or tens of meters in diameter and show a dust tail. More than 4500 of these objects have been cataloged in 25 years[12] thanks to the contribution of many amateurs and easy and almost immediate access to databases containing the images. Most of the objects, including those of a size much greater than 10 m, have been

---

[11] Michels (1982).

[12] A complete directory of grazing comets is maintained in the USA. The Sungrazer project, https://sungrazer.nrl.navy.mil/.

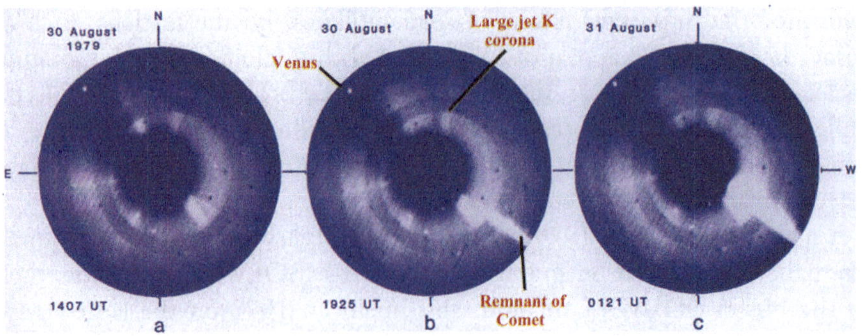

Fig. 8.10 Explosion of a grazing comet, from the Kreutz family, during its passage in the corona near the Sun. The video sequence is obtained by the externally occulted coronagraph of the US Navy, carried by the P-78-1 satellite and subsequently called SOLWIND. Such an explosion, never observed before, results in intense light at 45° down to the right. The comet, nevertheless deceased, was named after its discoverers at the Naval Research Laboratory NRL, Comet Howard–Koomen–Michels. (Source: Naval Research Laboratory)

identified as debris from a hypothetical giant comet more than 200 km in diameter, dating from several centuries or even several millennia. Work for historians?

These grazing comets belong to the family of Kreutz comets, named after the German astronomer Heinrich Kreutz who noted for the first time, at the end of the twentieth century, the similarities in orbit of several large comets that have been perfectly visible to the naked eye in the past, since the comet of the year 371 BC, observed with the naked eye by Aristotle.[13] Kreutz identified several of these historical comets, seen in broad daylight, as belonging to the same family. This family of comets represents about three-quarters of the SOHO objects listed. Among the large comets of this family that have recently approached the surface of the Sun and survived, we can mention the comets Ikeya–Seki 1965 and Lovejoy 2011.

For the first time, the passage of the comet Lovejoy 2011 through the corona was followed when it emerged from behind the solar disk, seen by the X-ray imager and far-ultraviolet EUV of the *Solar Dynamics Observatory* mission (Fig. 8.11). This instrument is a high-performance imager, called AIA, collecting more than ten very detailed images, with different filters in the X–EUV domain, every 10 s of time and continuously 24 h a day, during the SDO mission.

The analysis of the images shows surprising interactions of the nucleus of the comet, and especially the swirling debris that detaches from it (Fig. 8.12),

---

[13] Kreutz family, in English "Kreutz Sungrazer."

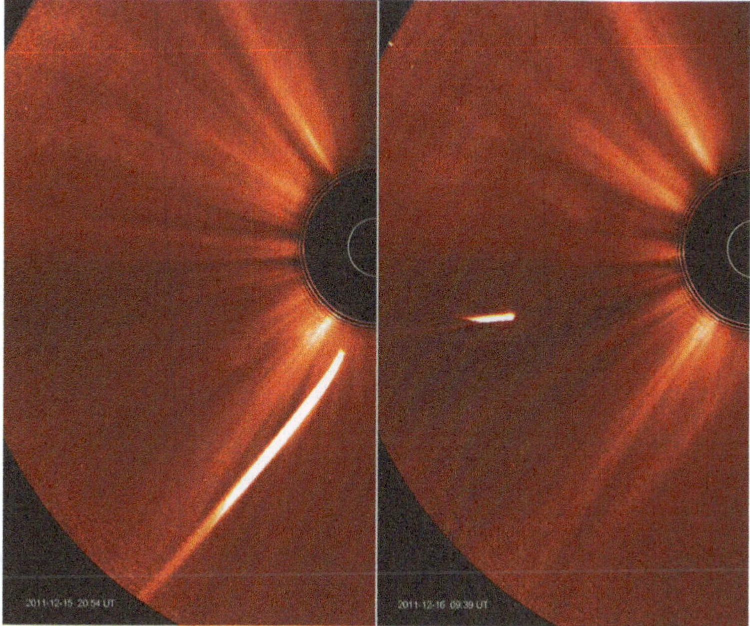

Fig. 8.11 Images of the tail of comet Lovejoy 2011, from the family of Kreutz comets, before (on the left) and after (on the right) its passage through the solar corona. Despite the intense heating received at a distance of only 0.2 solar radius from the surface of the Sun, this comet, with a nucleus diameter estimated at 200 m, has survived, but it disintegrated somewhat and its nucleus disappeared because it was completely sublimated. On these images made in visible light, the passage of the comet at its perihelion is, alas, obscured by the external occulter. Observations from the COR 2-A SECCHI coronagraph of the STEREO mission. (Source: ESA/NASA)

with the magnetic structures of the inner corona whose temperature exceeds one million Kelvin. This comet served as a probe for a detailed analysis of coronal physics by studying the fine structures of the plasma. Unfortunately, no observation in visible light or infrared was possible on board the SDO probe due to the lack of a coronagraph, so the production of enormous quantities of dust, which must occur at the time of the disintegration of the nucleus, could not be seen directly.

The dive into the inner corona and the reappearance of the comet were also observed by the EUV imager (known as SWAP) of the European microsatellite PROBA-2, launched in 2009. After its very turbulent passage through the inner corona, the comet still showed a beautiful, very extensive, and bright dusty tail, in images made over several days on the ground, notably in Australia, and even in near-Earth orbit, from the *International Space Station* (ISS).

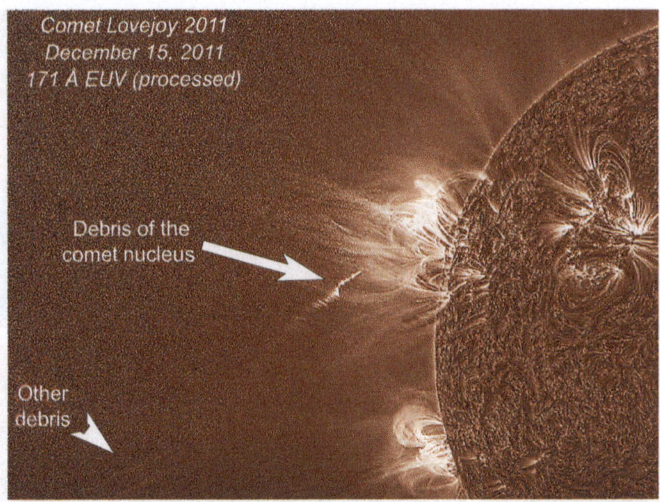

Fig. 8.12 Image of the debris of comet Lovejoy 2011 during the dive of the nucleus, which disintegrates in the inner corona with the pieces interacting with the magnetic field. This is visualized by the coronal structures that emit spectral lines in the extreme ultraviolet, around 17.1 nm. AIA instrument of SDO and image processed for a time-lapse film by M. Druckmüller (Source: Brno Institute of Technology/NASA)

The disintegration of the nuclei of small comets in the inner corona contributes to the supply of grains to the interplanetary medium, as demonstrated by Comet SOHO 2875 which, while not being a grazing comet, was discovered on SOHO images during a quite close passage in the corona (Fig. 8.13). The comet nonetheless disintegrated and was reduced, after its exit from the corona, to a long trail moving in the zodiacal cloud. Thus, in addition to the classic phenomenon of Poynting–Robertson slowdown on grains resulting from collisions in the asteroid belts, this cloud is also fed by small comets passing through the solar corona and undergoing the vicissitudes of the Sun's radiation, energetic particles, as well as the magnetic field emerging from the Sun and carried by the solar wind.

The total mass of the zodiacal cloud, dominated by the contribution of the F corona, is estimated to surpass by about two orders of magnitude that of the K corona of plasma, and being of the same order of magnitude as the mass of

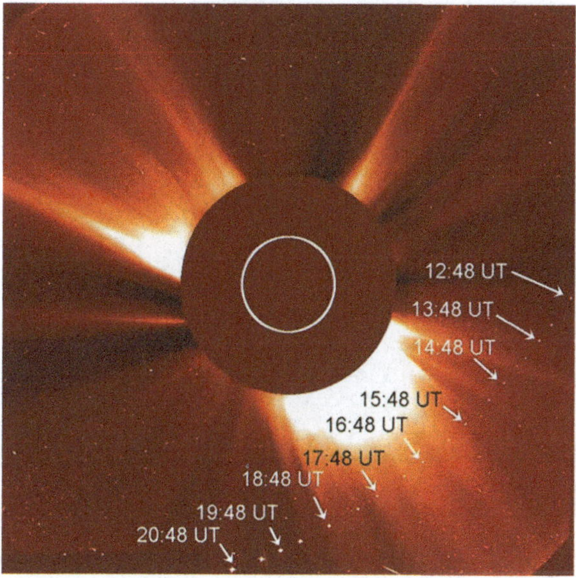

Fig. 8.13 Successive positions of comet SOHO 2875, observed during its passage on February 19, 2015. The comet, probably too small, had not been detected on the ground before its discovery with the SOHO coronagraphs, but its disintegration in the corona made it more visible after this passage. It is not a grazing comet from the Kreutz family. (Source: ESA/NASA)

the solar wind.[14] Nevertheless, for a more complete picture, it is important to note that the total mass flow (mainly ionized hydrogen and helium) of the solar wind corresponds to a mass renewal of the entire K corona about once a week that is four times during the time it takes for the Sun to rotate on itself. As for the F corona, the balances of quantities of solid matter brought in, coming from asteroids and comets, and of matter lost, evaporated near the Sun or blown away from it by radiation pressure, are still subject to speculation (Fig. 8.14).

---

[14] Kral (2017).

Fig. 8.14 The dust trail (elongated cloud), dispersed in the zodiacal cloud by the very small comet SOHO 2875. The dust is pushed by radiation pressure and moves away from the Sun somewhat like the dusty tails of normal-sized comets, which have a nucleus about 10 km in diameter. The comet lost its nucleus after its passage near the Sun. Image by the famous Austrian comet hunter Michael Jäger, see C/2015 D1 SOHO LRGB. (Source: https://spaceweathergallery.com/)

8.4   F Corona and Zodiacal Cloud: Pyramids in the Sky under Construction?

Having thus confirmed the role of comets in enriching the zodiacal cloud with dust grains, the images collected by the probe *Parker Solar Probe*, whose line of sight can approach down to ten solar radii from the surface of the solar disk, allow for a deeper understanding of these dust particles that form the F corona and the zodiacal cloud.

These two side-by-side images (Fig. 8.15) show two planets, Mercury and Jupiter, while the field of other sequences contains the Earth and Venus. The stable contributions, over the course of a sequence, from the F corona and the Z (for Zodiacal) cloud are subtracted, leaving only the images of moving objects. Thus, weak structures of the K corona, expanding almost radially, are highlighted. Numerous straight traces of meteoroids, quickly passing through the field, appear. At times, huge puffs or sprays of straight dust traces fill the image, but this parasitic phenomenon is not shown here. They are caused by the impact of a meteoroid on the spacecraft's heat shield, which would then lose a small part of its coating.

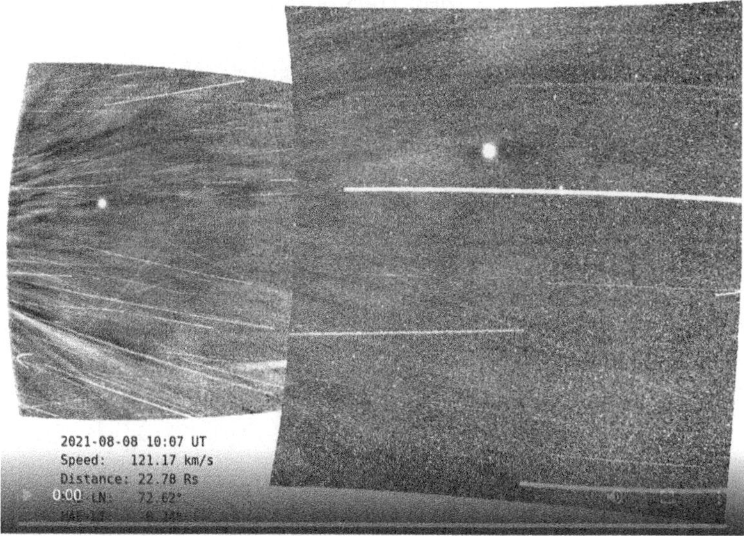

Fig. 8.15 Meteoroid trails on the wide-field images of the WISPR cameras of the *Parker* (NASA) probe in August 2021. These two images, extracted from long sequences, have been digitally processed to retain only "what moves" during a sequence. The field covered is, from right to left, over 100°. The Sun, out of the field, would be 10° to the left and hidden by a heat shield and occulting masks. For more images and movies at various resolutions, see https://wispr.nrl.navy.mil/encounter9-summary. (Source: NASA)

Discovery of Large Grains

The two wide-field cameras (WISPR), operating in visible light on board the *Parker Solar Probe,* have confirmed the existence of large grains, called alpha (α) meteoroids, in orbit near the Sun and belonging to swarms, remnants, or debris left by defunct comets.

In 2020, during several passes of the probe at its perihelion, one of these swarms was already been identified for certain, even if the signature of these large grains, with an estimated diameter of 0.5 mm, is not very distinct in the background of the images (Fig. 8.16). It would only be a faint component belonging to the well-known Geminids swarm, which are meteors (or shooting stars), visible each year around mid-December from Earth during our passage near their orbit. In confirmation of this interpretation, this swarm could be simultaneously observed from the ground in the Canary Islands, approximately with the same orbits.

It is laborious to highlight the trace of the swarms during the probe's passes, when it is closest to the Sun while traveling along its elliptical orbit. Indeed,

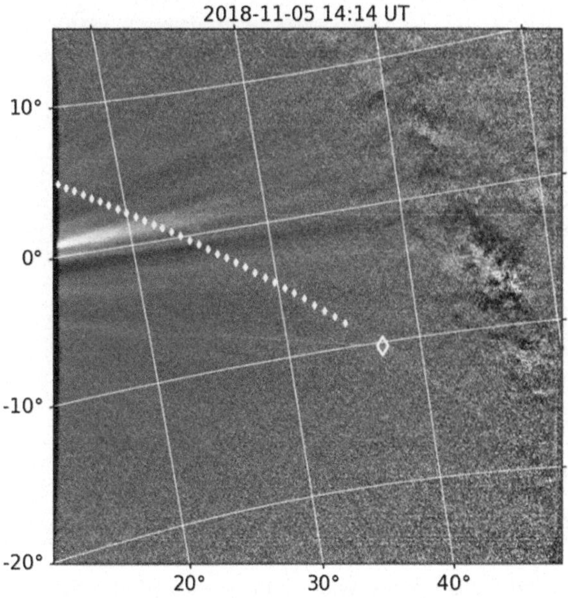

Fig. 8.16 Highlighting the orbit of a swarm of small grains near the Sun, by the imager of the *Parker Solar Probe*. A sequence of images, cleaned of stars, of the Milky Way (remaining on the right) and most of the K coronal jets (remaining on the left), is shown over a field of more than 30°, the Sun (on the left, out of field) being masked by an occulter whose vertical edge can be guessed at on the left. The diamond indicates the direction of the perihelion of the swarm's orbit (Battams (2020), https://arxiv.org/abs/1912.08838v1). (Source: courtesy of K. Battams, reproduced by permission of AAS)

it is first necessary to eliminate the light from the stars, giving a bright background, then that due to the K corona. Finally, α-type meteoroids or larger sometimes hit the probe's heat shield, and they produce debris that in turn scatters the sunlight and momentarily dazzle the cameras.

A selection of raw images is therefore necessary to avoid the most significant parasitic sprays. The high speed of the probe's movement, about 80 km/s in this region and its change in attitude due to its rotation are variables to take into account for a good determination of the trajectory of the swarms during the few hours that the observation lasts. It is remarkable to note that, so close to the solar furnace, these cameras are capable of detecting one hundred thousandth of a billionth (10^{-14}) of the brightness of the solar disk seen from Earth, which is about the background of the night sky on Earth on a moonless night and far from any city! These observations confirm that interplanetary grains, of millimeter size, penetrate into the circumsolar region and that their orbit can be determined.

The Nano-grains

The observations of the PSP probe also allow speculation on much smaller grains, such as nano-grains or β-meteoroids, smaller than a micron, present in the circumsolar region, whose orbital signature on the images, which is very weak, is hardly accessible to the cameras of the WISPR probe. In addition, they can be destroyed or at least ejected during their passage near the Sun, causing their possible sublimation (i.e., evaporation). The physics that governs them is the subject of several theoretical works[15] that emphasize the production, by erosion and collisions, of grains smaller than 0.5 micron. These grains were the subject of one of the instruments (Paris Observatory & Kitt Peak National Observatory) on board Concorde-001 during the 1973 eclipse (Part I).

The influence of this region on the more distant interplanetary medium and the role of the Sun's magnetic activity have long been the focus of the work of I. Mann[16] (Max-Planck Institute in Germany and University of Tromsö in Norway) who had first considered the collisions occurring within the zodiacal cloud as being the source of the nano-grains. With the data from the PSP probe, it becomes evident that it is the circumsolar region that is likely the source of these grains.[17]

Another question continues to intrigue, namely the relative concentration of the F corona around the solar Equator and further away, the ellipticity of the zodiacal cloud and its axis of symmetry close to the Ecliptic or the very close invariable plane of the solar system, which had been more or less explained by attributing the origin of the grains to collisions in the asteroid belt.

Revisiting the roles of the four forces mentioned above, and especially that of the force related to the electric charge of the grains, it turns out that the latter would play a dominant role on the smallest grains, less than 0.1 micron in size. The grains trapped by the solar wind would then tend to concentrate near the plane of the ecliptic, hence reproducing the in well-known concentration of the solar wind near this plane, itself quite close to the solar equatorial plane. But this still remains a hypothesis (Fig. 8.17).

Furthermore, note that the most numerous Sun-grazing comets (Kreutz family) are on orbits almost perpendicular to the plane of the Ecliptic and that, therefore, to obtain a concentration around the solar Equator, this

---

[15] Scazay (2021).

[16] Mann (2016), https://royalsocietypublishing.org/doi/10.1098/rsta.2016.0254.

[17] Mann and Czechowski (2021).

Fig. 8.17 Comet Neowise 2020 is among the brightest of the early twenty-first century. It is observed here by the WISPR camera of the PSP probe (NASA), far from its perihelion and against the background of the zodiacal cloud. The comet was classified as almost new with an orbit inclined about 50° to the ecliptic and a period estimated at 7000 years. Its dust tail is opposite to the direction of the Sun, located to the left behind the occulting mask and the thermal shield of the probe. At the time of this image, taken on July 5, 2020, the comet was approximately at the same distance from the Sun as the planet Mercury. The diameter of its nucleus is estimated at about 5 km. In circumnavigating the Sun at a respectable distance, since its perihelion was around 0.3 astronomical units from the Sun, a huge amount of dust and gas, including water vapor and sodium, was released into the interplanetary medium. The zodiacal cloud is faintly visible in the image, with its more or less pyramidal light cone. (Source: NASA)

electromagnetic force must be exerted radially during the production of the nano-grains. Parallel to this remark to explain the formation of an F corona concentrated around the solar equatorial plane, as well as the similar concentration of the zodiacal cloud, another possibility must be mentioned, long evoked in connection with the production of cometary grains more or less concentrated around the plane of symmetry of the solar system: the distribution of the orbits of periodic comets, with orbits less inclined than those of new comets which are on hyperbolic orbits or on very elongated elliptical orbits.

If the interplanetary grains of the zodiacal cloud came from the dusty tails of periodic comets, a concentration around the invariable plane of the solar system would seem to follow, with the additional effect of the gravitational interaction of the planets orbiting in the same direction.

This was studied by the French astronomers Philippe Lamy and Antoine Llebaria, specialists in the F corona and SOHO images, who measured the flattening (ellipticity) of the F corona for two solar cycles, or 22 years, with the LASCO coronagraph of the SOHO mission.[18] The remarkable observations of the *Parker Solar Probe* and the European *Solar Orbiter* mission reveal a circumsolar region and a zodiacal cloud much more active than what had been assumed to date. The time seems to have come for observations at wavelengths at which the thermal radiation of the grains is manifested, that is to say in the near infrared at less than ten solar radii from the surface, as was attempted on *Concorde* 001 in 1973 during a total eclipse. The cameras in the near infrared have made tremendous progress in their number of pixels, sensitivity, and miniaturization during the past 50 years as well as the computer resources enabling the processing of the images. A few seconds of observation during a total eclipse could therefore be enough. Favorable eclipses are expected in 2024 over North America and later over Europe. Other observations, focusing on the polarization or the spectrum of the F corona can still be considered as had been proposed to examine what has been called the solar Beta Pictoris effect,[19] named after a debris cloud around a star that our reader will find discussed in the following chapter. These results will obviously influence the study of exoplanet systems and especially the so-called habitable planets, as will be discussed in the next chapter.

Regarding the F corona and in a more distant future, one can dream of a large externally occulted coronagraph, observing in the infrared and operating in space, for example, from a space station as had been proposed[20] from the space shuttle, which has since been decommissioned. It would be the ultimate instrument for studying all of the dynamic phenomena of the circumsolar region.

---

[18] Lamy and Gilardy (2022).

[19] Shestakova (2003).

[20] The pinhole/occulter facility - NASA Technical Reports Server. https://ntrs.nasa.gov/citations/19830046550.

9

Debris Discs, ExoEarths, and Coronagraphy

Astronomers all remember the year 1995 as it marked the discovery of the first extrasolar planet, orbiting a Sun-like star named 51 Pegasi and located 50.45 light years from Earth in the constellation of Pegasus. By observing this object, two astronomers from the Geneva Observatory, Michel Mayor and Didier Queloz, thus ended a difficult multi-century quest and, above all, opened up a tremendous field of discovery of "extrasolar" planets, which we now call exoplanets. A quarter of a century later, thousands of planetary systems have been identified in our neighborhood—a "suburb of the Sun" extending a few hundred light years from it within our Galaxy. In 2015, the Nobel Prize in Physics rewarded these two discoverers.

But why devote our final chapter to this subject in a book dedicated to the story of the 1973 solar eclipse, and then to the current state of the solar corona?

We have recounted how, during the eclipse expedition on *Concorde* 001, the team from the Paris Observatory, associated with that of the Kitt Peak National Observatory in Arizona, chose to observe the F corona very close to the Sun, particularly the composition and temperature of the dust grains in the zodiacal cloud. These grains, deposited by cometary nuclei and spiraling toward the Sun, evaporate in the immediate vicinity of the star, perhaps forming the rings that the observation sought to detect during the eclipse (Fig. 9.1). Through the study of the F corona, this work, and many subsequent ones, has confirmed the existence of these tiny dust grains which, around the Sun, form this F corona that extends into the zodiacal cloud. Over the past half-century, we have learned much about their properties and origin, from the solar light that they scatter to the infrared radiation they emit, as shown in the preceding chapter.

P. Léna, S. Koutchmy, *Eclipsed Suns, the Solar Corona and Exoplanets*, Astronomers' Universe, https://doi.org/10.1007/978-3-031-92199-5_9

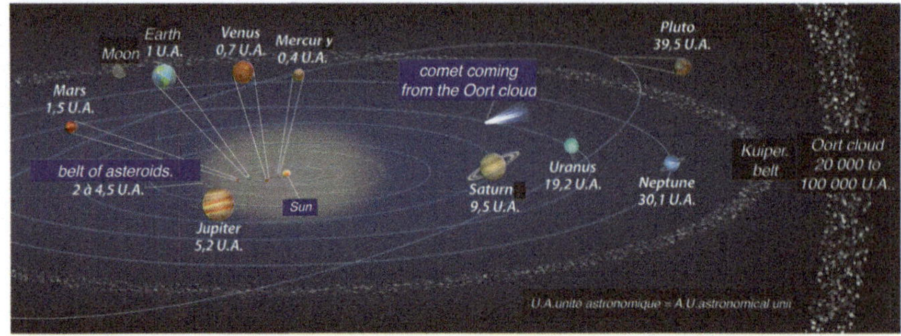

Fig. 9.1 Overall scheme of the solar system. Orbiting around the Sun are the terrestrial planets immersed in the diffuse F corona, the asteroid belt, the gas planets, and then the Kuiper belt and much further away, the Oort cloud. The intrusion of comets is suggested. The distances, in astronomical units (AU), and the dimensions are not at all to scale

In addition, the existence of the Kuiper belt (also known as Edgeworth–Kuiper), where larger grains and a quantity of small icy bodies and cometary nuclei orbit far from the Sun, has been confirmed as an extension of this zodiacal cloud, beyond the orbit of the planet Neptune, which is more than 30 astronomical units (AU) from the Sun.

All exoplanetary systems, discovered since 1995, have, to varying degrees, dust disks, which surround the star and play a crucial role in the formation of exoplanets. By hiding the dazzling light of the star and allowing an artificial eclipse, coronagraphs, derived from the one invented by Bernard Lyot, are now valuable tools for observing these discs and discern planets in them,[1] including by placing these instruments onboard observatories navigating in space.

Although, so far the survey has only reached a few thousand stars in the near neighborhood of the Sun, a solid statistical conclusion now leads us to believe that a significant proportion of the stars in the sky, at least those belonging to our Galaxy, are surrounded by these exoplanets. Moreover, it is more than likely that this is also the case in the billions of known galaxies. Thanks to a wide variety of observational tools, deployed or invented since 1995, the exoplanets that we already know have proven to be extremely diverse in their masses, their orbits around their host star, their history since their formation, the molecules present in their atmospheres. Our solar system, with its eight planets—Pluto having been stripped of the title of planet—and the beautiful regularity of their nearly circular orbits, seems a very particular case

---

[1] The rise of coronagraphy for the study of exoplanets is discussed in detail by Galicher and Mazoyer (2023). https://comptes-rendus.academie-sciences.fr/physique/articles/10.5802/crphys.133/.

within this recently discovered diversity, which includes objects 30 times more massive than our Jupiter (one thousandth of the mass of the Sun) to small planets, of size and mass comparable to those of our Earth (one three-hundredth of the mass of Jupiter).

Finally, among the multitude of questions raised by these objects, one of them can only attract even more the attention of researchers and humanity as a whole. Does our Earth have in the universe true twins that resemble it—we will call them exoEarths—with oceans and continents, atmosphere and, the ultimate question, the possible appearance and development of life? However, a too bright presence of zodiacal clouds in a distant planetary system makes particularly difficult this search for exoEarths, so small and therefore not very bright. Can we nevertheless hope for an answer to this question? In the USA and Europe, this question of "life in the universe" mobilizes considerable resources for the benefit of future space programs, both in the solar system (Mars and beyond) and in our close galactic environment. In Russia and in China, the goal of "colonization" of the near solar system (Moon, Mars) seems to be a priority, perhaps accompanied by a certain skepticism about the issue of "life elsewhere."

The extraordinary vitality around these questions, as well as the likely interest of our readers, and their close link with what precedes, justify that we address them here, only mentioning a few of the multitude of works, briefly and in a simplified way.

Before proceeding, here is the current definition given in 2014 by the International Astronomical Union, of what astronomers call an "exoplanet." These are "objects whose mass is low enough that nuclear fusion of deuterium (isotope of hydrogen) cannot occur—that is, less than 13 times the mass of Jupiter in the case of a star with chemical composition identical to that of the Sun [...]. These objects, regardless of how they were formed, are in orbit around stars or remnants of stars."[2] A long definition, therefore, which attempts to group in this category of "planet" the extraordinary diversity of objects discovered, whose masses vary by a factor of ten thousand, between the objects of mass comparable to that of the Earth, which we will call exoEarths and which are of special interest to us here, and those thirteen times more massive than the planet Jupiter.

Much less massive objects, such as "small planets" or "dwarf planets," exist in the solar system, and therefore very likely in other planetary systems. As they are still undetectable at these distances, they will not be discussed in what follows.

---

[2] For a full definition, published in 2022, see: https://www.iau.org/science/scientific_bodies/commissions/F2/info/documents/.

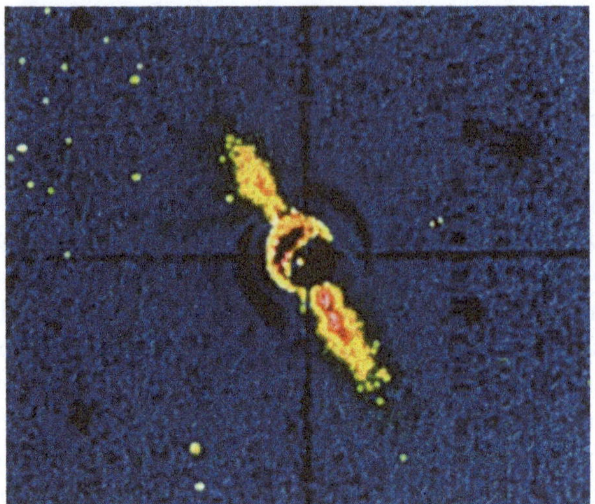

Fig. 9.2 The discovery of circumstellar disks. Around the bright star Beta Pictoris, visible to the naked eye in the southern hemisphere, the IRAS space observatory discovered in 1983 the presence of a dust disk, one among the "fabulous four," presented later (including Vega, Fomalhaut, and Epsilon Eridani). The following year, this historic image was obtained in visible light at the Las Campanas Observatory (Chile) by a 2.5-m telescope, equipped with a CCD detector and a Lyot coronagraph. The light of the disk is here the scattering of the star's light, this one being masked thanks to the coronagraph. (Source: courtesy of R. Terrile with the permission of *The American Association for the Advancement of Science*)

Launched in 1983 by NASA, the IRAS space observatory (*Infrared Astronomical Satellite*) was the first infrared observatory placed in space. Finally, infrared astronomy, whose airborne birth we have discussed in the first part of the work, took off, and discoveries began to pour in. The IRAS telescope, of relatively small size, did not make images of circumstellar disks, as its spatial resolution was insufficient. However, in a small number of cases, the simple measurement of the spectrum of the small region, including the star and its vicinity, detected an excess of light compared to what could be attributed to the star. Knowledge of the zodiacal cloud of the Sun allowed the immediate attribution of this excess to solid grains, heated by the star's radiation and located in its vicinity. Among the few cases detected in this way, the star β Pictoris stood out. A year later, its disk was observed in Chile with a terrestrial telescope (Fig. 9.2) by the astronomers Bradford Smith (Arizona) and Richard Terrile (California).[3] Forty years later, this star and its surrounding disk remain highly studied.[4]

---

[3] Smith and Terrile (1984).

[4] https://www.circumstellardisks.org/imgs/betaPic_SmithTerrile1984.png and https://w.astro.berkeley.edu/~kalas/disksite/pages/bpic.html.

9.1 Protoplanetary Disks

Within a galaxy, a star forms from an interstellar cloud, an immense volume of gas—mainly hydrogen and helium, with traces of carbon dioxide, water, methane—to which are mixed tiny grains of dust, most often silicates, of the same nature as the sand on our beaches. These grains were previously formed in the atmosphere of other stars at the end of their life, then ejected into space. Their size is around a thousandth of a millimeter (a micron), a dimension close to the wavelength of visible light or the near-infrared. Within this cloud, the force of gravity, mutual between these atoms and grains, will slowly act and grains will get closer, progressively forming a central condensation. The condensation attracts more grains and grows, this phenomenon being called accretion. As is most often the case, this cloud is initially slowly rotating on itself, so its contraction is more easily done in a direction parallel to its axis of rotation. The contraction results in the formation of a central dense core, surrounded by a flattened disk of gas and grains, perpendicular to the axis of rotation. The fate of the core is to quickly form the star, more or less massive, while that of the disk is to become a nursery for the formation of exoplanets, which is reflected in its name of "protoplanetary disk." This process is rapid, on astronomical timescales, taking place in a hundred thousand years.

As for our solar system, this general scenario of formation, conjectured by the French astronomer Pierre-Simon de Laplace as early as 1795, is now well established. Young stars and disks therefore coexist, as was verified as early as the 1980s on a well-known class of such nearby stars, called T Tauris because that star is the prototype and is found, like many others of the same type, in clouds of the constellation of Taurus. Later, it became clear that these disks were indeed the place of formation of planets, and they were then called protoplanetary disks. The phase intermediate is naturally designated as a "transition disk."

Figure 9.3 is a beautiful example of a protoplanetary disk. The disk is seen almost face-on from Earth. The light, at millimeter wavelengths, is emitted by the cool grains, at a temperature of a few tens of Kelvin because they are far from the star. Each gap, clearly visible on the image, reflects an absence of matter, which has perhaps served to form a planet, which has remained in orbit around the star but is not visible on the image because it is too faint. As early as 1975, the existence of this disk had been suspected from its radiation in the near-infrared, where the grains scatter the light of the central star by the same process as that of the solar F corona, this one observed in visible light.

Fig. 9.3 A beautiful example of a protoplanetary disk. Image of a disk around the star HL Tauri, observed in 2014 by the ALMA interferometer—an array of radio telescopes located in Chile—at millimeter wavelengths. The star is located 450 light years from the Sun and the radius of the disk is about 25 astronomical units, therefore less than the distance separating the Sun from the planet Neptune. The star emits practically no light at these wavelengths, so no coronagraph is necessary, and the star does not appear on the image. (Source: ALMA ESO/NAOJ/NRAO (ALMA Partnership. Vlahakis C., Rubens V.F., ESO Press release 1436, Nov. 2014))

A protoplanetary disk, initially very dense, will evolve and most often disappear. Its gaseous component has contracted into forming the star as well as massive planets themselves gaseous, like Jupiter or Saturn. Most of the grains have contributed, by agglomerating with each other, to form rocky planets, such as Earth or Mars, or smaller bodies, notably comet cores, which remain in orbit around the star at varying distances. The grains that have survived are "blown" toward the outside of the disk by the pressure exerted on them by the intense light of the young star (radiation pressure) or by the pressure of the "stellar wind," formed of electrons, nuclei, and atoms ejected in the case of less massive stars and colliding with the grains. We have encountered these two processes at work in the solar F corona. The oldest "old" disk known is only 25 million years old.[5]

---

[5] https://en.wikipedia.org/wiki/Protoplanetary_disk.

At the end of this process, as described here in broad terms, there remain in orbit around the star, more or less numerous planets, giant and gaseous or less massive and rocky, and a multitude of small solid bodies, called planetesimals (or planetoids), of kilometer size, also in orbit and which have not been able to grow to form a planet. The giant and gaseous planets form very quickly, in a few million years, while for the rocky planets and especially the exoEarths, the process can be longer.

9.2   Debris Disks

Debris disks represent a later phase of the evolution of a planetary system and they can persist for billions of years. As their name suggests, they are made up of solid bodies, ranging from tiny dust grains (one thousandth of a millimeter: a micron) to kilometer-sized bodies, both of which include not only rocky materials (silicates) but also water ices, carbon dioxide, methane, or ammonia. Unlike protoplanetary disks where gas is abundant, they contain little or no gas, as what remained has been "blown away." They coexist with already formed planets, as is the case in the solar system. However, the difference between these two stages—protoplanetary and debris—although convenient, may not always be so clear, apart from the amount of gas. Some debris disks, such as the emblematic disk around Beta Pictoris, are only a few million years old, barely older than the age of a protoplanetary disk.

Debris disk, but debris of what? The multitude of solid bodies with a rocky character, moving rapidly around the star and subject to the gravitational force of the already formed planets, leads to a large number of collisions between these bodies, breaking them into smaller pieces, which in turn collide with each other and break up. A cascade of collisions develops, leading to the size distribution that is observed in the debris disk, as different sizes do not emit or scatter light in the same way. But this distribution can only last a few million years, as forces other than gravity act on these small grains: the smallest particles are expelled outward by stellar radiation pressure, or fall in a spiral toward the star and vaporize, as observed in the solar F corona, or are destroyed by other phenomena. A factor called β (beta), greater than one for the smallest grains, thus measures the ratio between the force due to the radiation pressure of light and the force of gravity. These grains can also acquire an electric charge, by the photoelectric effect due to the ultraviolet radiation of the star.

A debris disk must therefore constantly renew itself to persist, and collisions are the mechanism that partly produces this renewal. Another source of renewal is the evaporation of comet nuclei, made of "dirty snow"—a mix of rocks and ices—and coming from very far from the star, when they approach it, as observed in the zodiacal cloud of the solar system. The case of the disk of the star Beta Pictoris is emblematic, since it was possible to directly observe the evaporation of comets in the immediate vicinity of the star, as discovered in 1986 by the French astronomers Alfred Vidal-Madjar, Anne-Marie Lagrange, and the team at the Paris Institute of Astrophysics. The gas produced by this evaporation causes a sudden and brief increase in the radiation emitted by the star. Its spectrum then reveals characteristic lines of cometary materials.

Close and Hot Debris, Distant and Cold Debris

Debris disks are extremely common. An estimate, made in 2018, indicates that 20% of stars similar to the Sun or more massive would have such a disk, more or less significant in mass.[6] These disks manifest themselves to the observer by two distinct components, which recall the zodiacal cloud and the Kuiper belt observed in the solar system.

The first component (Fig. 9.4) is located near the star, at a distance less than a few astronomical units. The grains, heated by stellar radiation, reach a temperature above 150–200 K. This somewhat arbitrary boundary is closer to the star the more massive it is because its luminosity, and therefore the ability of its radiation to heat the grains, increases very rapidly as the mass increases. Near the star, the temperature of the grains can, as in the solar F corona, reach more than 1000 K and the grains then vaporize. The infrared radiation emitted by this component is significant, and it allows us to determine the temperature of the disk, which is therefore "warm" near the star and cooler beyond. In addition, the grains, mostly the smallest and regardless of their temperature, scatter the visible light of the star intensely. By analogy with the solar system and the word "zodiacal", this part of the debris disk is often referred to as "exozodi." Making an image of this exozodi region, so close to the star, is difficult and we will return to it later.

The second component of the disk, also in orbit around the star, is further away, at distances of several tens, or even hundreds, of astronomical units from the star. The temperature of the bodies that make it up is therefore lower, less than 150 K, hence its name of "cold component." Here, collisions also occur but are less frequent because the space for movement is larger and the speeds of the grains in orbit around the star are slower. The presence of already formed planets is possible there, as observed around the star Fomalhaut by the *Hubble Space Telescope* (Fig. 9.6).

---

[6] Kral et al. (2017).

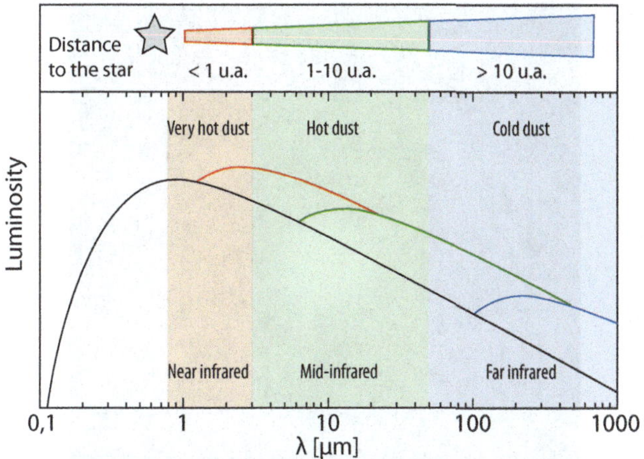

Fig. 9.4 Layout showing the intensity of light radiated by a star, comparable to the Sun and surrounded by a debris disk, as a function of wavelength, from the visible (0.3–0.7 microns) to the near-infrared observable from the ground (0.7–20 microns), then far- and sub-millimeter (up to 1 mm = 1000 microns). (Source: after Kral et al.)

Images of Disks

Ground or space telescopes, coronagraphs and adaptive optics, infrared wavelength interferometers with the European *Very Large Telescope* or in the millimeter wavelength range with the ALMA network, all these instruments have been contributing since the early 2000s to the observation of these disks, rich in indications on the formation of exoplanets.[7]

Indeed, these observational techniques significantly improve the angular resolution of large telescopes, and they provide otherwise inaccessible details.

Let us limit ourselves here to the presentation of some of these disks, illustrating this richness with images taken at different wavelengths and at different scales.

The Disk Around the Star Vega

Vega is one of the three beautiful stars of our summer skies, in the northern hemisphere. Its disk, discovered from space by the IRAS observatory in 1983, was observed in 2005 in the mid-infrared, at the wavelength of 24 microns, by the *Spitzer* observatory telescope, placed in space by NASA, which gives an image (Fig. 9.5). The disk is seen face-on from Earth, its axis of rotation being

---

[7] The site https://webdisks.jpl.nasa.gov/ contains a remarkable catalog of known disks, with many images when they are available.

Fig. 9.5 Debris disk around the bright star Vega, located 25 light years from the Sun, observed in the mid-infrared by the *Spitzer* observatory (NASA). The field is 2 × 2 arc minutes. (Source: NASA/Su et al.)

directed toward the observer.[8] The extension of the cold component reaches a distance to the star of 330 astronomical units (AU), but colder outer regions are observed up to 800 AU. In 2011, the exozodi component of this debris disk was also observed by its infrared emission, at a distance to the star of 8 AU.[9]

The Disk Around the Star Fomalhaut A

This bright star, visible in the constellation of Piscis Austrinus, is close as it is located 25 light years from the Sun, like Vega (Fig. 9.6). With a mass and temperature comparable to those of the Sun, but younger than it because it is only 440 million years old, it belongs to a triple system (A, B, C).

Three images of the disk gradually reveal its structure:

- Image a, taken at sub-millimeter wavelengths (0.45 mm), was obtained by the *James Clerk Maxwell* telescope from Hawaii by the British astronomer W. Holland in 2003.

---

[8] Su et al. (2005).

[9] Defrère et al. (2011).

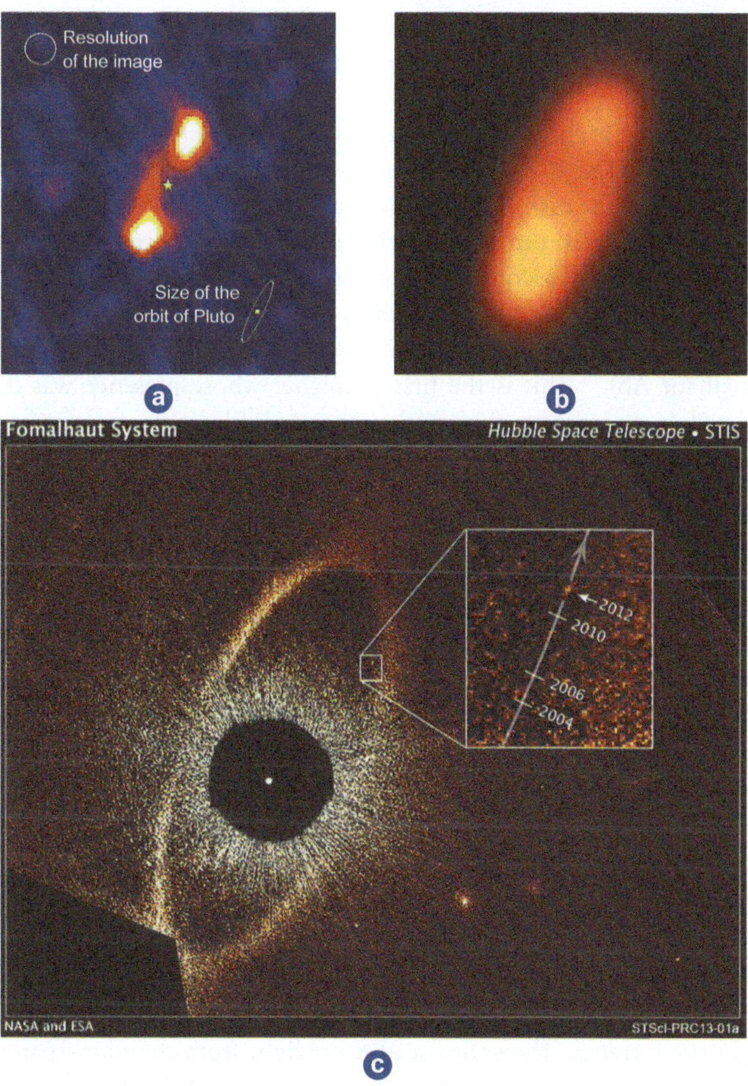

Fig. 9.6   Debris disk around the star Fomalhaut A. Observations of the cold, extended disk, compared to the orbit of Pluto in the solar system: (**a**) at sub-millimeter wavelengths from Hawaii; (**b**) in far infrared from space by the *Spitzer* mission; (**c**) observation of the visible light of the star, scattered by the disk, the star being masked by the STIS coronagraph of the *Hubble* telescope. A planet is identified, and its movement (insert) is tracked over time. The field is approximately 300 × 300 AU. (Sources: (**a**) East Asian Observatory (EAO/JCMT); (**b**) NASA/JPL-Caltech/K. Stapelfeldt (JPL); (**c**) *Hubble Space Telescope*/NASA)

- Image b, taken in the same spectral domain, but from space, by the telescope of NASA's *Spitzer* observatory (2005), more clearly shows the entire cold disk.
- Image c, taken at the wavelengths of visible light by the STIS coronagraph on the *Hubble Space Telescope*, is better resolved than the previous ones. The small grains of the disk, seen obliquely and with a radius of 100 astronomical units (a hundred times the Earth–Sun distance), scatter the light of the star, hidden by the disk of the coronagraph. The successive images, obtained between 2004 and 2012, reveal the presence of a planet, whose orbital movement can be tracked in the insert of the image (planet named Fomalhaut Ab).[10] This is the first exoplanet whose presence was directly detected on an image at the wavelengths of visible light, after the first exoplanet, directly identified on an image, was in 2004 in the near-infrared by the astrophysicist Anne-Marie Lagrange and her student Gaël Chauvin.

The Disk Around the Star Beta Pictoris

This star (in the southern constellation of the Painter), located 63 light years away from Earth, is hotter than the Sun with a surface temperature of 8000 K, but it is only 20 million years old (Fig. 9.7). The following four images show the progress in the observations of its disk.[11]

- The image a, dated 1987, was obtained in visible light by the coronagraph of the *Hubble Space Telescope*. The light is scattered by the grains of the disk. The disk extends up to 200 AU from the star.
- The image b was obtained with the European *Very Large Telescope* (the NACO instrument) in the near-infrared (3.6 microns). The quality of the image results from the use of an adaptive optics and a coronagraph operating in the infrared. The grains scatter the light from the star, regardless of their distance to it, the closest ones can also contribute by thermal emission.
- The image c was taken in the far infrared, around 0.1-mm wavelength, by the telescope of the *Herschel* space observatory (2010). The radiation is that of the low-temperature grains (a few tens of Kelvin) of the cold disk.

Inside the circle, the contrast between the background and the planet has been increased, to better highlight the latter.

---

[10] Kalas et al. (2013).

[11] https://webdisks.jpl.nasa.gov/show.php?id=16&name=beta%20Pictoris#refid1600.

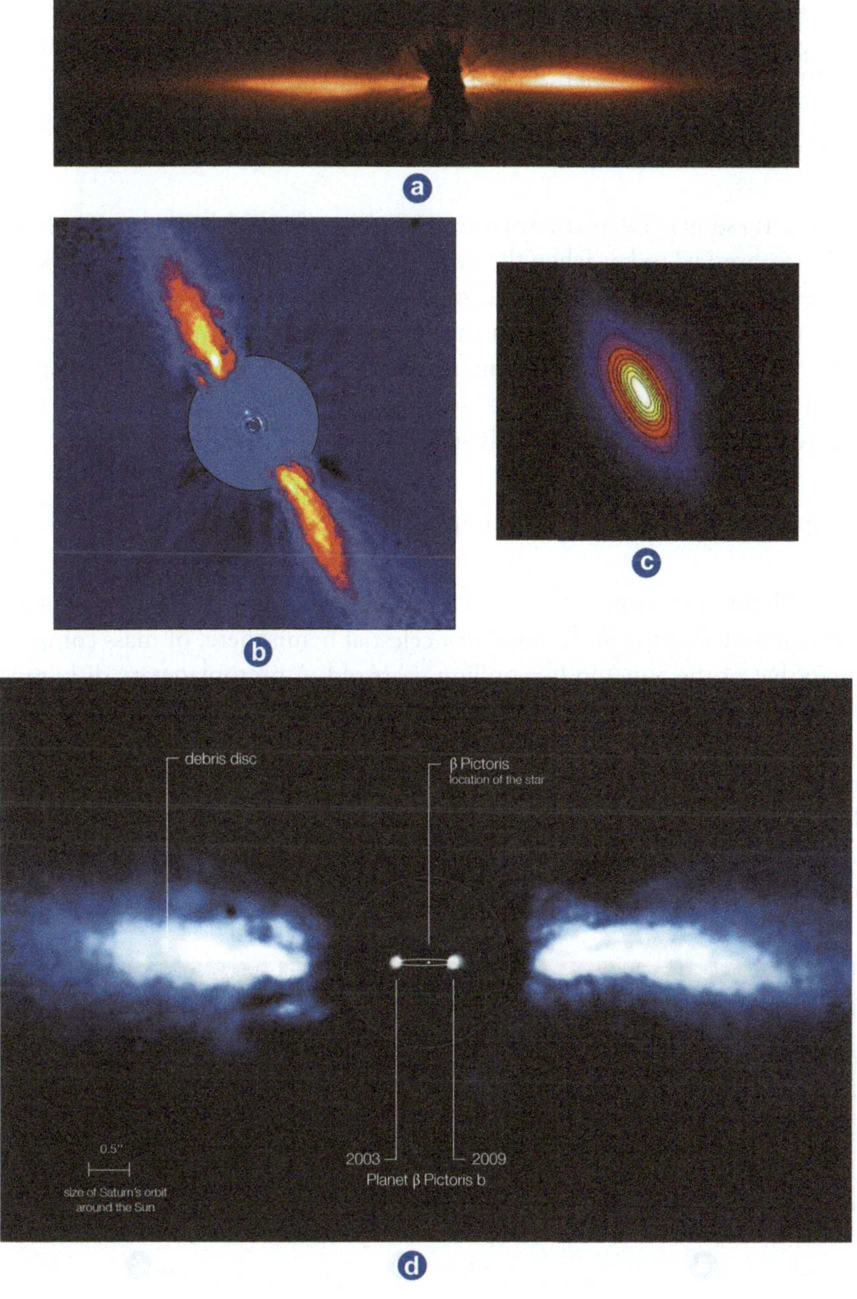

Fig. 9.7   Debris disk around the star Beta Pictoris. After the discovery of the debris disk in 1984 (Fig. 9.2), it is observed: (**a**) in visible light by the coronagraph of the *Hubble Space Telescope* in 1987; (**b**) in 1996 by the European telescope (ESO) of 3.6 m, and in 2008 by the *Very Large Telescope* European, from Chile, with a coronagraph equipped with adaptive optics in the near-infrared, showing a planet close to the star; (**c**) in 2010 by the European *Herschel* observatory from space at 100 microns wavelength; (**d**) in 2014, the coronagraph of the VLT confirms the movement of the planet on either side of the star. (Sources: NASA/ HST; ESO/A.M. Lagrange; ESA/Herschel (Vandenbussche et al. 2010); ESO/A.M. Lagrange)

- The image d reveals the movement of the exoplanet (β Pic b), whose mass is greater than that of Jupiter, orbiting the star at a distance of 10 AU (2008). A second planet (β Pic c), of comparable mass, orbiting at 2.7 AU, was discovered in 2020, in the exozodi component of the disk, which was also observed.

Were these massive planets, close to the star, formed in the region where they are observed today, when the matter present in this region of the disk was scarce? Or, formed at a greater distance, did they migrate toward the star, not far from its habitable zone, by a mechanism of friction with the disk, as is likely the case for the massive planets discovered even closer to their star and called "hot Jupiters"? Note that the search for traces of life does not favor these massive planets, which are entirely gaseous.

The Disk Around the Star PDS70

At 350 light years from us, the star PDS70 is a young star in the Centaurus constellation, located in the southern celestial hemisphere, of mass comparable to that of the Sun and five million years old. A protoplanetary disk, more precisely a transition disk, seen from three-quarters view, was discovered around this star. The coronagraph of the SPHERE instrument, built by a team around the French astronomer Anne-Marie Lagrange and installed on the European *Very Large Telescope*, provided an image in the near-infrared, in 2018. This image revealed the presence of a first planet, PDS70 a, massive and separated from the star by 25 AU (Fig. 9.8a).

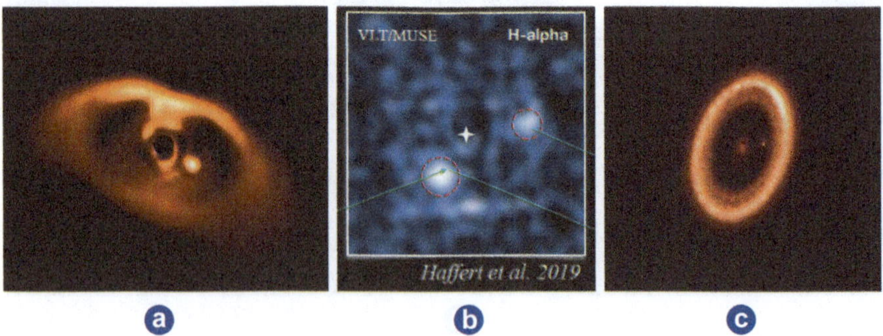

Fig. 9.8 A young disk, aged five million years and forming planets. Sources: **(a)** Keppler et al. (2018). ESO/A. Müller et al. (www.sci.news/astronomy/pds-70b-exoplanet-direct-image-06158.html). **(b)** Haffert et al. (2019), ESO (Haffert et al. 2019). **(c)** Benisty et al. (2021), ALMA (Benisty et al. 2021)

A year later, a second instrument called MUSE, on the same VLT and also designed by a Frenchman, the astronomer Roland Bacon, detected a second planet PDS70 c, of comparable mass and slightly more distant, by the light emission of a cloud of hydrogen surrounding these planets (Fig. 9.8b). The same year, the ALMA interferometer, installed in Chile and observing millimetric radio radiation, provided a third image where the disc, which is cold because it is far from the star at the distance of a "Kuiper belt," appears perfectly, as well as the two planets (Fig. 9.8c). The star at these wavelengths does not radiate much and therefore does not represent a disturbance that would dazzle the image.

This protoplanetary disc is particularly remarkable, since we "see" both the presence of a dust disc and the formation of two giant planets, comparable to Jupiter.

These few examples have illustrated our brief journey toward the discs, whether they are protoplanetary, debris, transition, or exozodiacal, to the extent that these categories are really understood and represent the diversity of these objects as new as fascinating. It is clear that, most often, the formation of a star is accompanied by the formation of a disc of gas and dust. It is clear that this disc, which is subject to the action of the star's radiation, evolves over time, particularly with the appearance of planets more or less massive which, in return, interact with it. In addition to the formation of planets, it is clear that the dust grains of the disc have their own history: forming small rocky bodies and comets, which in the solar system are present in the Oort cloud and the Kuiper belt. Finally, to conclude this point, it is clear that if we have learned a great deal about discs since the beginning of the twenty-first century, many questions remain before we can accurately place our own solar system in the vast family of exoplanetary systems, tomorrow within our Galaxy and probably much later by having a broader vision including other galaxies.

9.3   Detecting exoEarths

The zoo of exoplanets is rich, with more than 5000 exoplanets at the beginning of 2023 and more than a thousand multiple-planetary systems, where the star is surrounded, as in our solar system, by several planets. However, only about 30 of these thousands of exoplanets have a mass between half and twice that of the Earth, of which only about ten are less than 50 light years from the Sun. These objects are in orbit around spectral type M stars, less massive and cooler than the Sun. However, it should be noted that these M-type cool stars are often active, ejecting particles into space that can have a destructive impact on the complex molecules of living matter.

This observed rarity of planets comparable to Earth certainly does not reflect reality, but simply the difficulty of detecting these small mass or small-size objects. The transit method is the best for highlighting their presence around a star. When, from Earth, the planet's orbital plane is seen edge-on, the planet passes in front of its star each period, producing a partial eclipse of the star's light that we call a "transit." Although the decrease in the star's brightness is weak, the smaller the planet is small, this method, implemented with ground-based telescopes, but especially those placed space since the early 2000s, is extremely fruitful, particularly for larger planets. We can thus mention the French mission CoRoT (2006–2014), the *Kepler* missions (2009–2018), TESS (since 2018) from NASA, and CHEOPS (since 2019) from the European Space Agency.

Everything indicates, however, that debris disks must form a large number of planets similar to Earth or even less massive. Even if a precise estimate of the number of these objects in the close vicinity of their star remains difficult, the key is to have a sensitive enough method to discover them, count them, and form their image without being blinded by the star's brightness.

Observed at a distance not too distant from the Sun, for example, 50 light years, by a powerful telescope that forms an image of it, how would our solar system appear, based on what we know about the F corona, as well as its extensions in zodiacal light and the Kuiper belt? The Sun, very bright particularly in visible light, dominates, with a brightness that constantly decreases as one moves toward the infrared (Fig. 9.4). A faint luminous oval surrounds it, aligned with the ecliptic plane, whose total brightness is about one ten-millionth of the total luminosity of the Sun, corresponding to the dust disk with its two components—the zodiacal cloud and the Kuiper belt—being equally bright. Finally, close to the Sun, the three rocky planets, three luminous points including Earth, which appears ten times less bright than the zodiacal component. Let us observe the Earth from this distance of 50 light years. Let us put ourselves in the most favorable case when it is seen in quadrature in its orbit, that is to say during its "first quarter." It is then at the tiny angular distance of 60 thousandths of a second of arc from the Sun, drowned in the emission of the zodiacal cloud that extends between 20 (0.3 AU) and 200 milli-arcseconds (3 AU). The goal of the observation, made in the near-infrared, will be to separate and therefore clearly distinguish in the image, the Earth from the zodiacal cloud. To not be prevented by diffraction, which limits the angular resolution power of any telescope, a large telescope whose main mirror must measure at least 10 m in diameter is necessary.[12] In the

---

[12] The progress made in recent decades in the ability of large terrestrial telescopes to discern extremely fine details in the images that they form is described in Léna (2020).

image provided by such an instrument, it would then become possible to distinguish the luminous point standing out on the diffuse background of the zodiacal halo, and to identify the presence of the Earth. To achieve that, it would obviously have been necessary to eliminate the light of the Sun, ten million times brighter than the zodiacal light, using an advanced coronagraph (Fig. 9.9).

This simple description helps us to appreciate the immense difficulty that exists to discover the presence of exoEarths in this region close to the star where they were formed and are most likely present. We therefore understand that the systematic search for exoEarths around nearby stars requires determining the light of the exozodi. Indeed, known debris disks and the few observations available in 2022 indicate a considerable diversity in the intensity of exozodis: the detection of exoEarths will be all the more difficult as these exozodis will be bright, compared to what we can call "the exozodi of the solar system," which is now quite well known and can therefore serve as a reference.

To go further, an image is necessary, capable of separating in infrared light the contribution of the star from that of the exozodi. But here, unlike the beautiful images of disks presented above and obtained with a coronagraph, it is necessary

Fig. 9.9 Artist's rendering of the Trappist-1 planetary system, with its seven planets of masses comparable to that of Earth in the habitable zone and in orbit around a cool dwarf star located 50 light years from the Sun and gradually discovered by the team of Michael Gillion at the University of Liège (Belgium) from 2015 by the transit method, on the ground and in space (Grimm et al. 2018. See also https://www.eso.org/public/news/eso1805/). (Source: ESO & NASA/JPL)

to be able to distinguish details even closer to the star. So, since the beginning of the 2000s, interferometers combining several ground-based telescopes and offering resolutions comparable to those of a single telescope of several tens of meters in diameter have been used for direct images of the thermal radiation of exozodis close to their star: the European *Very Large Telescope* in Chile and the *Large Binocular Telescope* in Arizona. Thus, it has been possible to observe in a few cases lower infrared intensities, less than a hundred times the solar exozodi.

There is a second reason why the discovery of an exoEarth near the star arouses so much interest. This is the notion of a "habitable zone,"[13] namely the region of space located around the star where the average equilibrium temperature reached by a solid body, planet or otherwise, heated by the star's radiation, is between 0 °C and 100 °C, which allows liquid water to exist on its surface. Now on Earth, the development of life, at least three billion years ago, seems to have required liquid water, then present in the ocean(s). The importance of this notion of a habitable zone makes its definition still the subject of discussion. Let us, however, give the essentials.

The spherical shell, thus delimited in space around the star, has an inner radius (100 °C) and an outer radius (0 °C), both of which are larger the more massive, and thus luminous, is the star. With the Sun whose total radiation is equivalent to that of a sphere of temperature equal to 6000 °K, the inner and outer radii of the "habitable" shell are today 0.95 AU and 1.67 AU respectively, taking into account the presence of the Earth's atmosphere and its greenhouse effect. Indeed, the equilibrium temperature of the Earth, with its orbit and without its atmosphere, would be −18 °C, while with this effect, due mainly to carbon dioxide (CO_2) in the atmosphere, its average temperature reaches 15 °C. The Earth has become habitable, and even inhabited!

The exozodi component of the debris disk and the habitable zone associated with the star are two regions, close to each other and the star, which often overlap, which reinforces the interest of direct observations giving images of exozodi clouds, either on the ground or in space.

In 2028, the new giant European telescope, called the *Extremely Large Telescope* (ELT), will come into service in Chile (Fig. 9.10). It is equipped with a 39-m diameter mirror and equipped with extremely high-performance adaptive optics systems. Its coronagraph will then be able to detect the light emission from the dust disk of the order of one billionth of that of the star in the near-infrared less than 0.1 arcsecond from the star: this corresponds to a separation of 1 AU for a star located 30 light years from the Sun.

---

[13] en.wikipedia.org/wiki/Circumstellar_habitable_zone as well as https://en.wikipedia.org/wiki/List_of_potentially_habitable_exoplanets.

Fig. 9.10 Artist's conception of the *Extremely Large European Telescope*, whose main mirror measures 39 m in diameter, installed on the site of Cerro Armazones, in the Atacama Desert in Chile, not far from the *Very Large Telescope* (VLT/VLTI), which has been in service since 1998. The six laser beams are used for the operation of an adaptive optics system, which removes the degradations due to the Earth's atmosphere and allows one to obtain images of extremely high angular resolution. (Source: ESO/L. Calçadas)

9.4   Replace the Moon to Eclipse a Star

Since observations from the Earth's surface are so difficult, since the 2000s studies and proposals have multiplied, aiming to put observational tools in space to advance the discovery of exoEarths, within or close to the habitable zone of their host star. The first step in this quest is to determine the importance of the exozodi component around a large number of stars, which will be chosen as close as possible to Earth to be able to distinguish between the exozodi and the star as well as possible in an image. The observation will therefore first require making a detailed image in the near-infrared, using a space telescope whose primary mirror diameter should reach several meters. It will then be necessary to create an artificial eclipse, to best eliminate the star's light thanks to a coronagraph and detect the presence of the exozodi, and perhaps of planets present in this zodiacal cloud.

Almost all of the stars in the sky within 100 light years from the Sun have been cataloged, very completely,[14] by the three instruments operating in visible light and installed on the modest telescope (1.45 × 0.5 m) of the *Gaia*

---

[14] Smart et al. (2021). See also www.cosmos.esa.int/web/gaia/edr3-gcns.

space mission, launched by the European Space Agency in 2013. The catalog of observations made by *Gaia* covering our entire Galaxy includes 1.7 billion stars, a fraction of which have been precisely recorded in position and speed. Thus known, the total number of stars in this volume close to the Sun is about 8000, the majority being comparable to the Sun. There is also a less numerous population, which is difficult to detect because very dim, of very cool stars (spectral type M), around which it will also be interesting to search for exoEarths, since it is around such stars that the first exoEarths were discovered, such as those of *Trappist*-1 presented above.

After this inventory made by the *Gaia* mission, the systematic exploration of the environment of these few thousand stars, and therefore of their debris disks and notably of the exozodi component, is the objective chosen by ambitious exploratory projects in space, during the 2020 decade or the following ones, developed by NASA with numerous international cooperations, notably European (France and the European Space Agency). They will be very costly and will require new technologies and significant development times, which can reach two decades before the launch of a new observatory. For its space program called *Voyage 2050*, Europe is leaning toward a major mission aimed at the "Characterization of temperate planets," a mission whose study must be pursued, notably taking into account the projects of the USA,[15] who define their own objective: "The search for biosignatures on a robust sample of 25 planets, located in the habitable zone [of their star] ... This NASA mission has been named Habitable Worlds Observatory and the target period for the launch is the first half of the 2040s."[16] The mirror diameter chosen for the telescope capable of achieving this goal is 6 m, thus comparable to that of the *James Webb Space Telescope*, successfully launched in 2022.

Regardless of the detailed concepts that the studies on these objectives will lead to, one necessity remains: that of being able to distinguish small mass exoplanets, therefore not very bright, from the brightness of their star which they are necessarily close to, since they are located in its habitable zone. We present here two of these ideas and pre-studies: the *Roman Telescope*, to be launched in 2027 and the mission *Habitable Worlds Observatory*, further in the future, both of which illustrate the ambition of artificial eclipses for the detection of exoEarths.

---

[15] ESA (2021).

[16] NAS (2023), https://nap.nationalacademies.org/catalog/26522/origins-worlds-and-life-a-decadal-strategy-for-planetary-science.

The Roman Telescope and Its Coronagraph

The 2.4 m telescope of the *Roman* mission (NASA), initially called WFIRST (for *Wide Field InfraRed Survey Telescope*), will be launched in 2027 and placed near the Lagrange point L_2, 1.5 million kilometers from Earth in the direction away from the Sun.[17] It is named after Nancy Grace Roman (1925–2018), the first woman to hold an executive position at NASA and nicknamed the "mother of Hubble" for her major commitment in the realization of the *Hubble Space Telescope*.

The Lagrange point L_2 is particularly suitable for long-duration observations, and it already hosts the *James Webb Telescope*, the successor to the *Hubble* telescope and a marvel of robotic and computer technology since 2022. One of the two main objectives of the *Roman*/WFIRST mission is the study of the exozodi component of the debris disks around these nearby stars, using a high-performance coronagraph. Called the *CoronoGraphic Instrument* (CGI) and operating in visible light, it is based on the same principle as the classic Lyot coronagraph, but with spectacular innovations in active optics and occulting mask. It will be able to distinguish a point source of light, such as an exoplanet, a hundred million times less bright than its host star, at a distance of 0.15 arcsecond from it—this value corresponds to a separation between the source and the star of 1 AU if this star is located at 22 light years from the Earth. Regarding the detection of exozodi, the CGI instrument is designed by an impressive international team led by Frenchman Bertrand Mennesson, who works in California at the Jet Propulsion Laboratory (Fig. 9.11). This instrument will be able to detect a very weak exozodi, only three times more intense than that of the solar system, and thus access exoEarths that a too bright exozodi would hide.[18]

By the end of the 2020s, when the *Roman* mission will have completed the bulk of its observations, our understanding of exozodi will have advanced, answering current questions and certainly raising new ones: where do the grains that allow them to persist come from (destruction of comets, cascades of collisions between asteroids, …)? What are the processes that make these grains disappear (radiation pressure outward, spiraling in toward the star, evaporation)? What gravitational disturbances do the grains suffer from already formed planets? Can the light emission of the grains tell us about the presence of planets? Around which stars is this emission minimal? The choices

---

[17] https://roman.gsfc.nasa.gov/ and https://roman.ipac.caltech.edu/sims/Circumstellar_Disk_Sims.html.
[18] https://arxiv.org/pdf/1909.02161.pdf.

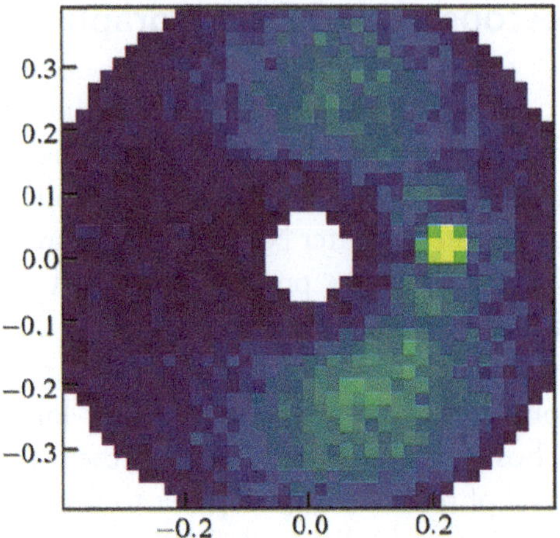

Fig. 9.11 Simulated image of the exozodi around a solar-type star, located 30 light years away. The zodiacal cloud is seen face-on in visible light, with a 2.8-h exposure, by the CGI coronagraph of NASA's *Roman Telescope*. The scale is in arcseconds. A hypothetical massive planet is located 1.6 AU from the star (on the right). (Source: B. Mennesson/JPL)

to be made for future missions, directly dedicated to the direct detection of exoEarths in our vicinity, will then be clarified.

A Disk Navigating in Space

Discerning exoplanets by eclipsing the star around which they orbit is the goal of upcoming missions. Therefore the NASA mission Habitable Worlds Observatory (HWO) is exploring different strategies to achieve this goal. Since an internally occulted coronagraph, even placed in space, still suffers from residual stray light, due to the star which can be a billion times brighter than the planet, can we achieve an artificial total eclipse, like total solar eclipses where the disk hiding the Sun is that of the Moon? It would therefore be necessary to place an opaque disk perfectly aligned with the axis of the telescope, far enough from it so that the disk eclipses the star but not the planets located around it. The problem may seem insoluble, but it does not seem out of reach to the engineers of the Goddard Space Flight Center (NASA), who accomplished the extraordinary performance of deploying the mirror in space in 2022, and since then operating the *James Webb Space Telescope*.

Let us examine this idea of an "artificial Moon" capable of producing an eclipse, which has been considered for the *HabEx* concept, studied by NASA since 2018 and now merged within the HWO.[19] Nevertheless, we present it

---

[19] https://www.jpl.nasa.gov/habex/.

Fig. 9.12 The *HabEx* concept, studied by NASA until 2022 to identify exoEarths around stars close to the Sun. The external disk of the coronagraph (on the left), which is jagged to modify its Arago spot, is placed thousands of kilometers away from the telescope (on the right). The occulting disk and telescope, in their respective orbits around the Sun, are slaved to each other, to maintain the necessary alignment with the star. In the direction of the disk, aimed at by the telescope, two exoplanets are represented, in orbit around the hidden star. The figure is obviously not to scale, but gives a general impression of the concept. (Source: NASA, op.cit)

here since it explores in detail the idea of a free-flying occulting disk, masking the stellar light and all allowing to see the tiny ExoEarths who may orbit around the star. It is necessary to design the opaque disk so that it ensures the maximum suppression of stellar light, while being certain that this disk remains exactly in the axis of a telescope located tens of thousands of kilometers away from it, and ensuring its deployment and the stability of its shape (Fig. 9.12).

The concept therefore considered a disk 50 m in diameter, placed 76,000 km from a 4-m diameter optical telescope that targets the star surrounded by a planetary system, assumed to be located less than 30 light years away. The disk is aligned with the axis of the telescope and the direction of the star, it has small motors to maintain this alignment. Elementary geometric considerations show that then the star is totally hidden by this disk, which therefore realizes a true artificial eclipse.

On the other hand, a planet, 1 AU distant from from the star, is seen by the telescope in a very slightly different direction and its light is not masked: the telescope forms its image at its focus on a camera, and the image is transmitted to Earth.

Here, before describing in more detail what this mission would be capable of observing, we need to take a detour to the nineteenth century, when the nature of light provoked lively debates among the scholars of the time.

Box 9.1 The Maraldi–Poisson–Arago Spot

Giacomo Filippo Maraldi, born in 1665 in the States of Savoy, arrived at the Paris Observatory in 1687. Jean-Dominique Cassini, director of this observatory, founded 21 years earlier by Louis XIV, had invited his young nephew Giacomo, who would go on to have a brilliant career as an astronomer,[20] to join him. In 1723, with his student Joseph Nicolas Delisle, Giacomo observed a strange phenomenon. He illuminated an opaque circular disk with a very small light source, then placed a screen at a distance, on which he obviously observed the circular dark shadow of the disk. To his great surprise, a bright point appeared, precisely in the center of the shadow.

This strange discovery would fall into oblivion for a century, while scholars debated the nature of light. The corpuscular conception that Newton had was then challenged by the new idea that light is actually a wave, able to propagate in a vacuum as acoustic waves do in air, as shown in 1802 by the British Thomas Young through a famous experiment. In France, the Academy of Sciences wondered, divided between the Newtonians and the anti-Newtonians, and put the question to the 1817 competition. The very young engineer Augustin Fresnel participated and submitted to the competition a memoir which will stand the test of time. He mathematically described the light wave as well as the diffraction phenomena that disturb it when it encounters an obstacle.21 To contradict Fresnel's results, a brilliant mathematician and pro-Newtonian, Siméon Denis Poisson, a member of the Academy, showed through a beautiful and entirely accurate demonstration that, if one follows Fresnel's ideas, a bright point should be observed at the center of the shadow of a screen illuminated by a bright point. This inevitable conclusion appeared obviously absurd to the proponents of Newton's corpuscular optics, of which Denis Poisson was a part: the disk must obviously stop all light particles.

François Arago conducted the experiment with a disk of 2.5 mm in diameter, the result was indisputable, the bright spot was indeed there: Fresnel won the competition and the wave nature of light becomes inescapable throughout the nineteenth century. The bright point, now called Arago's spot, is nothing more than a strict consequence of the diffraction of the light wave by the edges of an opaque disk.

In 2017, radio astronomers were able to observe Arago's spot at radio frequencies when a very distant galaxy called a quasar, forming a point light source, was found exactly occulted by the asteroid Palma, with a diameter of 190 km and then 3.4 AU away from Earth.[22]

---

[20] https://fr.wikipedia.org/wiki/Giacomo_Filippo_Maraldi_gallica.bnf.fr/ark:/12148/bpt6k3527h/f124. item. And (Hecht 2002).

[21] Aime et al. (2013).

[22] Harju (2017), https://iopscience.iop.org/article/10.3847/1538-3881/aad45b.

Regarding the *HabEx* concept, a problem remains, which is not insignificant: this Arago–Poisson–Maraldi spot, which will illuminate the telescope very precisely, would ruin all the effort of an artificial eclipse by its presence. It is therefore necessary to get rid of it, by calling on the laws of Fresnel, that is to say by giving the edge of the opaque disk a different outline. Perhaps, by choosing it well, we can use the unavoidable diffraction of light on this new contour, to make disappear or at least reduce this spot's brightness, by forcing the light to go elsewhere? The calculation as the experiments carried out at the Jet Propulsion Laboratory (California) lead to choose for the edge of the disk a splendid form of flower, with 24 petals of 16 m long, forming a corona grafted onto a central circle of 20 m in diameter (Fig. 9.13). This light assembly is made of thin carbon-fiber composite. Folded in the launcher's fairing on leaving Earth, the screen would unfold in space, thanks to a technique already used for radio antennas. The screen will have to be placed 76,000 km from the telescope, on an orbit chosen so that the alignment star–disk–telescope is maintained at all times, the latter tracing its own orbit (Fig. 9.13). When the alignment is very precisely achieved, the shadow of the screen covers the telescope and hides the star.

The diffraction of its light, observed in the near-infrared, by the petals of the disk greatly attenuates the Arago spot, which is then only one ten-thousandth as bright as the residual contours (Fig. 9.13, on the right). The

Fig. 9.13 (left) The screen, with a diameter of 52 m, studied for the NASA *HabEx* concept. The central ring (blue) is made of solar panels. (right) 76,000 km away, the shadow of the screen, as drawn by the diffraction of light in the near-infrared, when a star is exactly in the observation axis. The color code corresponds to the residues of starlight in the shadow cast at the telescope on a square about 50 m on each side. The intensity of the light in the spot at the center (blue) measuring 3 m is about one ten-thousandth of that of the edge (orange). This central light, residue of the Arago spot, is used for a feedback signal in order to maintain the disk–telescope alignment. (Source: Extract from the *HabEx*/NASA report, 2021)

4-m diameter telescope, centered on the spot, and the image detector at its focus, therefore receive a very weak residue of starlight. This residue determines the minimum brightness of an exoEarth that can be detected. Moreover, it permits constantly controlling and correcting the alignment of the screen with the axis of the telescope, thanks to the small thruster motors at the screen.

The telescope and the screen would therefore be launched separately and travel together for 6–8 months to reach their "parking" at the Lagrange L_2 point, located 1.5 million km from Earth, from which their main mission will take place, which is expected to last 5 years. During this long journey, which was also that of the *James Webb Space Telescope* after its launch on Christmas Day 2021, the screen is deployed, the instruments of the telescope gradually activated and checked.

Pointing the telescope toward the star on the one hand, and aligning the screen on the other hand thanks to small gas jets, are maneuvers well known now. Although tens of thousands of kilometers separate the two systems, they experience almost exactly the same gravitational force. This mainly combines the effect of the Earth and that of the Sun, constantly aligned for the instruments located at L_2, the effect due to the Moon being much weaker: this facilitates the control of the mutual position of the two systems. The most critical point is the exact maintenance of the center of the screen in the axis of the telescope, to within a few decimeters in lateral movement, in order to ensure that there is no leakage of the intense light of the star toward the telescope. The Arago spot, measured by infrared sensors (Fig. 9.13), indicates any deviation from alignment, which allows a correction to be sent to the disk motors 76,000 km away.

Four complementary but distinct objectives can be assigned to the observation program. Are there exoEarths in orbit in the habitable zone of certain nearby stars? Are there exoEarths whose atmosphere contains water vapor? Are there exoEarths showing signs of life? Are there exoEarths with water oceans (H_2O)? We have already highlighted the existence of other methods of studying disks and exoplanets, and the first question will undoubtedly have found elements of an answer using spectroscopy which indirectly detects the exoplanet by its gravitational effect on the star, as Michel Mayor had done. As for the second question, it may also be addressed by observing transits, when the light from the star, having passed through the planetary atmosphere at that time, retains traces of the atmospheric components of the planet.

The desired performance is then the following, in visible light or in the very near-infrared: to be able to observe at a distance of 0.1 arcsecond from the star, with the ability to distinguish a planet, located in the habitable zone of this star and whose brightness is greater than one ten-billionth of that of the star; then obtain spectra, which would indicate the composition of the atmosphere, looking for the signature of water, methane, oxygen, and ozone. The exploration perimeter extends out to distances of 30–60 light years from Earth and would include about 50 stars. Without going into the complex detail of the instruments envisaged at the focus of the 4-m telescope, the reader understands that this extraordinary performance requires, among other things, an exceptional mastery of coronagraphy. This is why the *HabEx* concept is studying two complementary configurations: The first includes a classic Lyot coronagraph, placed behind the telescope and with internal occultation, whose masks are improved, without therefore using the distant screen. The second configuration of external coronagraphy, called SHADE (shadow), uses this very remote screen and thus improves by nearly a factor of 3 the detection sensitivity.

The *Habitable Worlds Observatory* under study will certainly evolve over the next two decades before the instrument capable of meeting the challenges that we have briefly described is finally launched. Let us therefore bet on success, and let us conclude this chapter with an image, such as the *HabEX* mission might hope to achieve, or whatever it will have become in 20 years, and its coronagraph. This image is obviously a digital simulation of what the instrument "could see," and the simulation (Fig. 9.14) focuses on the detection of exoplanets around a star in the constellation of **Canes Venatici**, 27 light years away, with the performance of the *HabEx* concept using the SHADE external disk coronagraph. Five planets are detectable, including one exoEarth. The zodiacal cloud, from which this exoEarth stands out, is barely visible at this scale while a zone, analogous to the Kuiper belt in the solar system, is clearly visible, forming a ring seen obliquely, whose radius is comparable to the distance from Neptune to the Sun. The zodiacal cloud and outer zone here have an intensity five times greater than those of the solar system. The size of the image is 100 astronomical units on each side.

All of this probably seems very distant to our readers of the 2020s. We nevertheless like to point out this future when, beyond our Sun, whether observed in a natural eclipse or with a coronograph, other artificially "eclipsed Suns" will exist for astronomers, namely stars! Certainly, the hoped-for discovery of a nearby exoEarth, possibly showing signs of biological activity, could be considered one of the greatest scientific and human adventures of the twenty-first century, the outcome of which no one can yet predict.

As such, its preparation will take several decades, during which new results and technological progress will accumulate. These represent a necessary step, requiring tenacity and patience, until the launch of a mission in which researchers from all over the world will collaborate, seeking an answer to the question that each and every one of us can ask, faced with the mystery of our own existence on our small planet, Earth.

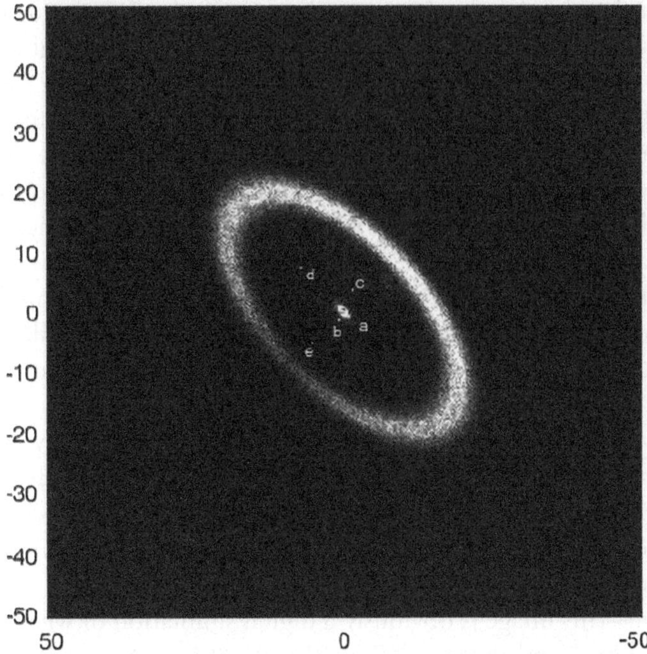

Fig. 9.14 Simulation of the detection of exoplanets around the star β CVn, 27 light years away, by the *HabEx* mission using the SHADE external disk coronagraph. The scales are in astronomical units (AU). (Source: *HabEx* report, NASA, op. cit)

9.5   Conclusion

Before you close this book, dear reader, let us take one last look at the journey that we have proposed. First, a story: that of a beautiful but ephemeral eclipse adventure, especially when compared to the immense effort of successive generations, seeking to understand our Sun. Through this story, perhaps we have managed to share the daily activity of the researcher, the difficulties and joys encountered, the indispensable collaborations. However, and intentionally, we have kept somewhat distant from the science, suggesting it rather than detailing and explaining it. Therefore, this story deserved that another facet of this science be highlighted, that of knowledge and ignorance, that of hypotheses and solid progress, that of the interdependence between astronomy and physics. This highlighting was all the more necessary as the 50 years that have passed since the flight of *Concorde* 001 have transformed our understanding of the Sun and opened up new perspectives, well beyond our solar system. Even if the coronagraph, this brilliant invention of Bernard Lyot in 1936, had provided an extraordinary instrument for the study of the solar corona outside of eclipses, the contribution of instruments carried on spacecraft in recent decades has transformed what we understand about this corona. The story of our first part therefore had to be followed by a more scientific look, undoubtedly more difficult to read but beautifully illustrated by the beauty of the images, a look that was the subject of the second part of the book.

The fabric of this book is thus imbued with the indelible emotion that the observation of a total solar eclipse leaves. This majestic and brief celestial phenomenon periodically marks the life of our civilizations, bringing a surprising lesson in astronomy. For us who are researchers, it was and remains a privileged moment of encounter with the corona of our Sun. It is then that the secret aspects of heliophysics are revealed. For a few minutes of wonder, the grandeur of the Universe overwhelms us, this Universe where our place is actually so modest! May this book prepare our reader to experience such a moment, on the ground or perhaps on an aircraft, or even from the deck of a ship, in August 2026 in Iceland or Spain, a year later in Egypt or in Saudi Arabia....

Unlike all other stars, our Sun is close enough to us that it can be studied in all its details, or almost. This simple luminous ball of hydrogen and helium, whose central nuclear cauldron transforms a bit of the hydrogen atoms into light, has gradually revealed more and more of its extreme complexity. A magnetic field, produced by electric currents circulating inside the Sun, then on the surface, configures the coronal plasma into beautiful loops, arches, and jets. It contributes to blowing the solar wind into interplanetary space, and it is the source of light and high-energy particles. We could not avoid presenting the role of this central character, despite the subtlety of the physics that governs it.

We represented our planet Earth as a simple ball of rock moving around the Sun in an interplanetary space almost empty of all matter. Gradually, this image has changed. This space has filled with rapidly moving electrically charged particles, with dust grains of all sizes brought by visiting or periodic comets, produced by collisions between a myriad of asteroids, which bear witness to the formation of the solar system. The events, which constantly occur on the surface of the Sun and in its corona, at the rhythm of the 11-year cycle of activity, modify interplanetary space and interact with the Earth, are so important for our daily life that a new term has been coined to designate these interactions: space weather.

Finally, for centuries, we have been questioning the existence of other worlds: are there other planetary systems around other stars? Would they be habitable? Does our Earth have twins in the Universe? During the past three decades, this age-old question has found the beginning of an answer.

Exoplanets exist in our Galaxy by the thousands, and even millions without much doubt. We wanted to show that this opening adventure of exoplanets is not without connection with the discoveries made on the Sun, its corona, its environment of dust and gas, the place of our Earth, nor with the observational techniques born around solar eclipses. Exploring these new worlds through observation and thought, without however turning away from the problems of our humanity, here is the perspective now open, capable of arousing enthusiasm and the vocations of researchers, of mobilizing the necessary resources from everyone.

Thus goes the edifice of science. How better to conclude than to resume here what Hamlet said to Horatio: *There are more things in heaven and earth, Horatio, than are dreamt of in your philosophy.*

Correction to: Eclipsed Suns, the Solar Corona and Exoplanets

Correction to:

P. Léna, S. Koutchmy, *Eclipsed Suns, the Solar Corona and Exoplanets*, Astronomers' Universe, https://doi.org/10.1007/978-3-031-92199-5

The original version of this book was inadvertently published with errors in Chaps. 1, 6, and 7.

- In Chap. 1: Eclipses and Humankind: On page 3, lines 2–5, a comment by the author on proofs that had erroneously been reproduced in the text has been dropped. The text suppressed is "Initially, Fig. 1.1 was conceived as a frontispiece for Chap. 1, without being referenced. In order to comply with rule to have every figure referenced in the text, I have to add a short sentence referring to the figure."

The updated version of these chapters can be found at
https://doi.org/10.1007/978-3-031-92199-5_1
https://doi.org/10.1007/978-3-031-92199-5_6
https://doi.org/10.1007/978-3-031-92199-5_7

- In Chap. 6: Seventy-Four Minutes of Observation, But What Gain for Science? Figure 6.4 appearing on page 96 was incorrect. The correct Fig. 6.4 is:

Fig. 6.4 This picture is taken from Skylab, at the time it passes close to the band of totality. The Moon (upper left) is not aligned with the Sun, which is in the center, covered with a coronographic Lyot mask held by a support (shadow at lower left). The Moon is illuminated by the Earth's shine and the corona is well visible, enhanced by a numerical treatment of the image. (Source: NASA)

- In Chap. 7: The Corona and Its Gaseous Component: In page 152, the caption of Fig. 7.29 was incorrect. The correct caption is:

Fig. 7.29 Image extracted from a two-dimensional numerical simulation of coronal evolution with time—along the solar limb in the photosphere and perpendicularly toward space above the solar surface. Unit of distances in abscissa is 3000 km. The distances are in thousands of kilometers. The simulation calculates the temperature (left) on a scale 0–4 MK and the density (right) on a scale 0–10 particles/cm^3 at each point, taking into account the magnetic field whose lines of force in white are sometimes open to the outside, sometimes closed in loops, sometimes in reconnection with ejection of a jet of plasmoids toward the outer corona. The small arrows indicate the movements of the plasma. The elapsed time (marked t) is 7.5 min after the initial stable state (calculation carried in 25 time steps of 18 s each). (Source: Yokohama and Shibataka 1998, reproduced by permission of American Astronomical Society)

Appendix 1: How do Total Solar Eclipses Occur?

Why Are Total Solar Eclipses So Rare in a Given Place on Earth?

To predict the occurrence and details of a solar eclipse (e.g. in Senegal, as in Fig. A.1), one must take into account the annual movement of the Earth around the Sun relative to the stars, the monthly movement of the Moon around the Earth, and finally the daily rotation of the Earth on itself.

These three movements, also called "revolutions", have different and independent periods, the first on the order of the year, the second of the month, the third of the day. The movement of the Earth around the Sun takes place in a plane called the "ecliptic plane", while the movement of the Moon around the Earth takes place in a different plane (Fig. A.2).

The orbits of the Earth around the Sun, on the one hand, of the Moon around the Earth on the other hand, are not circles but slightly oval curves called "ellipses". The distance between the Earth and the Sun, or the Earth and the Moon, therefore changes during a revolution traveled on each of these ellipses, of which the Sun or the Earth occupy one of the foci and not the center. The apparent diameters of these two celestial bodies, seen from Earth, are *almost* exactly the same, half a degree, which is a pure coincidence. This has not always been the case—in a very distant past and will not be in a similarly distant future, since the Moon is slowly moving away from the Earth due to the tides. In addition, the apparent diameter of these two celestial bodies, seen by a terrestrial observer, changes slightly depending on whether the

© Springer International Publishing Switzerland 2025
P. Léna, S. Koutchmy, *Eclipsed Suns, the Solar Corona and Exoplanets*, Astronomers' Universe,
https://doi.org/10.1007/978-3-031-92199-5

Fig. A.1 With these beautiful stamps from 1973, the postal services of the Republic of Senegal sought to explain to the entire world the mechanism of a solar eclipse, total or partial. In Senegal, which was not crossed by the band of totality, the eclipse was only partial

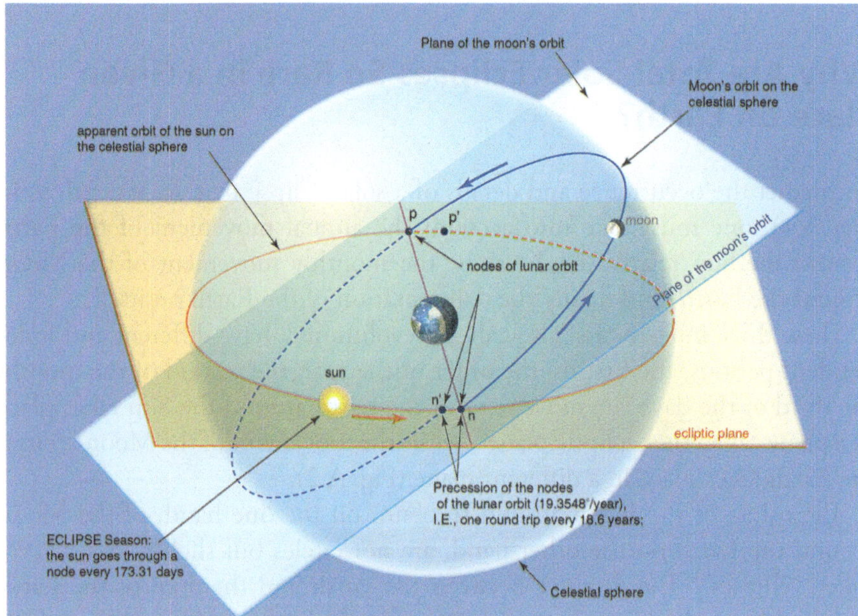

Fig. A.2 Positions of the ecliptic plane (yellow), of the lunar orbital plane (blue), relative to the Earth (at the center). Seen from Earth, the Sun (yellow) describes a trajectory (in red, Ecliptic), and the Moon (crescent) similarly describes a trajectory (in blue). The line **pn** is the line of nodes at a given moment, the points **n** and **p** moving over time to **n′** and **p′**. To make the figure more instructive, the angle formed by the ecliptic plane and the plane of the Moon's orbit has been increased to nearly 45°, while its actual value is about 5°

Moon is at its closest near the Earth (perigee) or not, and whether the Earth is closest to the Sun (perihelion) or not, since both of the orbits are slightly elliptical.

It is 29.531 days—lunation or synodic period of the Moon, that is 29 days 12 h 44 min and 3 s—that separate two consecutive Full Moons—when viewed from Earth the Moon is in the opposite direction to the Sun—or two New Moons—when Sun and Moon are in the same direction. This duration of the lunation results from the combination of the first two movements mentioned above.

One might expect that at each New Moon, the Moon passes between the Earth and the Sun, thus producing a total eclipse. This would be the case if the plane in which the Moon's orbit is located were the same as the plane of the Ecliptic. Each total eclipse would then be separated by 29.531 days. In fact, these two planes are distinct, there is a small angle between them: 5°8′43″. The Moon's orbit intersects the plane of the Ecliptic at two points diametrically opposed, which are called "nodes". The "line of nodes" is the line that joins them and is the intersection between the two planes. If, viewed from Earth, the line of nodes coincides with the direction of the Sun when the Moon passes through this node, then there is a solar eclipse. This eclipse is partial or total depending on whether the coincidence is more or less exact, producing a more or less large overlap between the two disks, as the overlap depends on the distances of the Earth to the Moon and to the Sun. Finally, the occurrence of a solar eclipse therefore depends on the orientation of the line of nodes relative to the Sun.

Celestial mechanics, which governs the three movements mentioned above, also governs the orientation of the line of nodes over time. Under the effect of gravity, the direction of this line slowly rotates relative to the direction of the Sun seen from the Earth, making a complete rotation in 18.61 years. Therefore, this fourth movement must be taken into account in predicting eclipses.

Figure A.3 summarizes the situation, viewed from the center of the Earth. The plane of the Ecliptic is projected onto the celestial sphere and traces a virtual line, called the "Ecliptic", cutting through the constellations (these are the signs of the zodiac). At a given moment, the direction of the center of the Sun is located on a point of the Ecliptic (longitude of the Sun). This point moves over the course of the year. It passes through one of the two nodes every 173.31 days (about twice a year).

Viewed from Earth, the plane of the lunar orbit is also projected onto the celestial sphere and traces another virtual line.

At a given moment, it is possible to plot on this line both the direction of the center of the Moon in longitude and latitude, and the direction of the nodes, which is always on the Ecliptic.

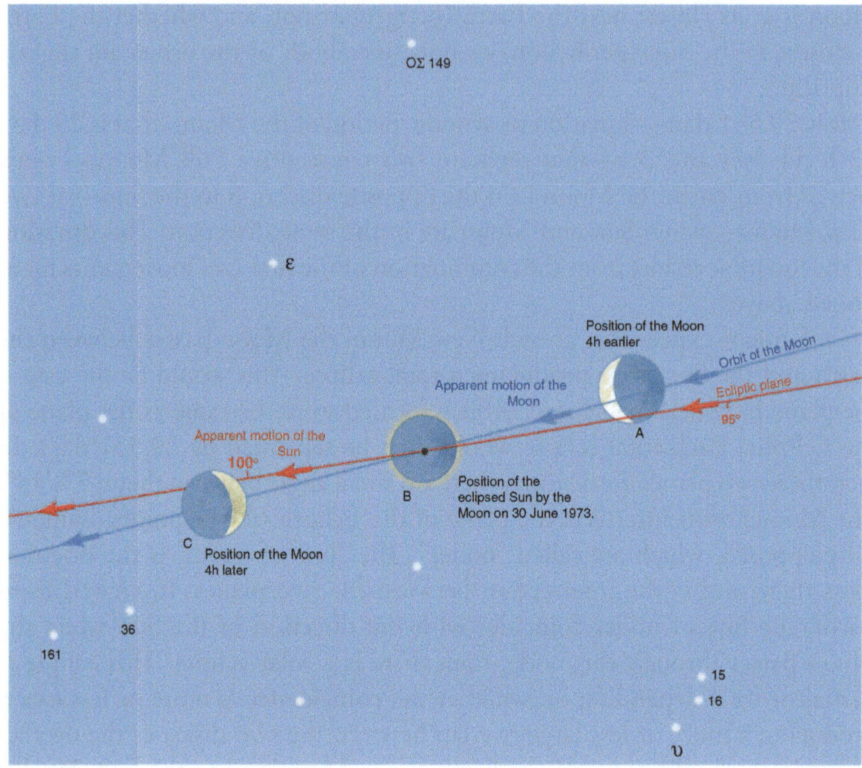

Fig. A.3 Appearance of the sky seen from Western Africa, on June 30, 1973 during totality when the stars become visible. The position of the eclipsed Sun (to scale) is indicated. It is located in the constellation of Gemini and coincides at this moment with the Moon, at an ecliptic longitude close to an angle of 100°. Over time and for a terrestrial observer, the Sun moves slowly on the Ecliptic (red line) relative to the stars. Similarly, the Moon moves relative to the stars (blue line), but about twelve times faster. Its trajectory is plotted, as well as its position 4 h before and 4 h after totality. The stars that are in a direction close to that of the Sun are indicated with their identification

The Moon orbits and passes through the same node (one of the two) during each orbit. The duration separating these two passages is 27.212 days (called the *draconitic period*). It differs from the duration of the lunation (29.531 days) seen above, as it is independent of the movement of the Earth around the Sun, which is not the case for lunar phases, such as the New Moon.

Finally, the Sun and the Moon, as seen from Earth, are not points but discs extended against the background of the stars; it is possible to plot the size of these discs around their respective centers.

From the above, it becomes possible to understand what determines the occurrence of solar eclipses, total or partial, and their repetition.

Let us refer to Fig. A.3. The exact moment of the New Moon is when the two longitudes, that of the Sun (which moves on the Ecliptic) and that of the Moon, are equal; this happens during each monthly lunar revolution. A solar eclipse can then occur around this moment, and we must then take into account the size (about half a degree) of the solar and lunar discs:

- If the latitude of the center of the Moon is less than 1.42°, the eclipse occurs for certain.
- If the latitude of the center of the Moon is between 1.42° and 1.58°, it may occur.
- If the latitude of the center of the Moon is greater than 1.58°, it does not occur.

It is possible to find the correspondence between these angles and the longitude of the center of the Sun, by comparison with the longitude of the node of the lunar orbit. Remembering that the angle between this plane of the lunar orbit and the plane of the Ecliptic is small (about 5°, see above), the above eclipse conditions correspond to larger angles for the difference of longitudes:

- If the difference between these longitudes is less than 15.665°, there is an eclipse.
- If the difference is between 15.665° and 17.375°, there may be an eclipse.
- If the difference is greater than 17.375°, there cannot be an eclipse.

Depending on the values of the differences in latitude or longitude at the passage of the Moon near the node, and depending on the apparent diameter of the two bodies at this moment, the solar eclipse will be partial or total. When total—it can be annular (apparent diameter of the Moon smaller than the one of the Sun), beaded (named also diamond ring, when the diameter match so exactly that light passes only between mountains at the Moon's edge), or fully total (apparent diameter of the Moon larger than the one of the Sun)—can the movements of the Earth around the Sun, of the Moon around the Earth, and of the line of nodes in the plane of the Ecliptic combine to ensure a certain regularity to the eclipses? We can observe an interesting coincidence: 242 times the draconitic period is exactly equal to 223 times the synodic period of the Moon. In other words, after $242 \times 27.2122208 = 223 \times 29.5305882 = 6585.32$ days, the centers of the Sun, the Earth, and the Moon are found to be very nearly in the same relative positions, and a new cycle begins. This duration of 6585.32117 days = 18 years 11 days 8 h is called the "saros". A saros contains on average 42 solar eclipses, of which 14 are partial and 28 are central (total, beaded, or annular). These numbers can vary slightly from saros to saros. The *exeligmos* is a cycle of three saros, designated by this Greek term meaning "revolution" or "orbit", but also referring in Greek to the sinuous curve that a fleeing hare follows! Two total eclipses, e.g., in a diamond ring configuration (when totality leaves solar light pass between mountains of the Moon), separated by three saros cycles occur at the same local time on Earth.[1] By convention, the numbering of saros begins in the year 2000 BC, thus covering all historically reported eclipses. The saros in which the eclipse of June 30, 1973 occurred was number 136.

As celestial movements are known with great precision and are reproduced identically over time, it is possible to calculate the date and time of each eclipse. Numerous tools are available on the Internet to provide this information, for example the very remarkable French site of the astronomer Xavier Jubier[2] which has an excellent page on the eclipse of 1973, or that of the Institute of Celestial Mechanics and Ephemeris Calculation, from which the above is taken.[3]

Why Is a Total Solar Eclipse Observed Only from a Very Small Part of the Earth's Surface?

When we observe the shadow of an object created by a light source, two cases can occur: the source is either a point of light, or of extended dimension. In the first case, the shadows are perfectly sharp, at the transition between dark and light areas. In the second case, we observe between total shadow and full light an intermediate area called the "penumbra", partially lit by a part of the light source. Since the Sun, as seen from Earth, is an extended source and not a point of light, the shadow of the Moon on the Earth presents a penumbral zone. Given the apparent size of the Sun being half a degree, the size of the Moon and its distance from Earth, the diameter of the shadow is at most 269 km when the Moon is closest to Earth (perigee) and that of the penumbra is 3000 km. The eclipse is therefore total when it is observed from a point on

---

[1] Robert Morris is perhaps the first to have been able to superimpose two images of totality taken at an exeligmos interval (three saros) in 1966. (Personal communication).

[2] Xavier Jubier, a French amateur astronomer, provides on his site a precise and user-friendly calculator of past and future eclipses. https://xjubier.free.fr/site_pages/Solar_Eclipses.html and https://xjubier.free.fr/site_pages/astronomy/ephemerides.html.

[3] Site of the Institute of Celestial Mechanics and Ephemeris Calculation (IMCCE, Paris Observatory): https://www.imcce.fr/fr/ephemerides/phenomenes/eclipses/soleil/chap01.php.

Earth located in the shadow, and partial if this point is located in the penumbra. Outside the penumbra, no eclipse occurs.

The movement of the shadow at the Earth position goes from West to East due to the direction of the Moon's movement in its orbit around the Earth, giving a speed of about 3400 km/h. But the rotation of the Earth on itself, also from West to East, subtracts from this speed the speed of Earth's rotation, which depends on the latitude of the place considered and is 1660 km/h at the Equator. The result of this difference in speed is the final resulting speed of the shadow on the ground, which is therefore between 3400 km/h near the poles and 1740 km/h at the equator.

In space, the region in which the Sun completely hides the Moon is a cone, the "shadow cone". The actual distance from the Moon to the Earth means that the tip of this cone gets at or very close to the Earth's surface, which allows the Moon to be able to completely and briefly hide the Sun. Spacecraft, put into space and moving in the Solar System, also experience a total solar eclipse when they cross this shadow cone.

The position of the band of totality on the Earth's surface is determined by the conditions of the eclipse (longitudes and latitudes), which in turn determine the point where the center of the shadow is on the Earth's surface and the area, called the band of totality, that the shadow will sweep.

Statistically speaking, for a given place on Earth, an eclipse will only be total every 270 years. Many places are inaccessible (ocean, poles) and the probability of a clear sky in an accessible place is also limited. That is why astronomers, after calculating the position of the totality band, always carefully choose their observation sites.

Why the Total Eclipse of 1973 Was Named the "Eclipse of the Century"?

A total solar eclipse reaches its maximum duration when the shadow spot is as large as possible, i.e., when the Earth is as far away from the Sun as possible (aphelion), the Moon as close to the Earth as possible (perigee), and the Sun is near the zenith of the observation place, in which case the shadow is circular. The first condition sets the date of such an eclipse, which in the northern hemisphere must be close to the summer solstice. The third sets the observation place, which must then be close to the Tropic of Cancer (again in the Northern hemisphere). Under these conditions, the speed of the shadow is 2196 km/h, its diameter is 262 km, which results in a duration of 7 min and

10 s. The calculation can be somewhat improved, and the duration at a point further South of the tropic or only 5° from the Earth's Equator can reach the absolute maximum of 7 min and 30 s, as the diameter of the shadow then decreases, but the speed, and the ratio of the two quantities, is more favorable. Other situations may provide shorter totalities (Fig. A.4).

For the annular eclipse, the conditions are a bit less strict and opposite: the apparent diameter of the solar disk must be the largest possible (Earth at perihelion) and that of the lunar disk the smallest possible (Moon at apogee). The duration of the annular eclipse can reach 12 min 30 s.

On June 30, 1973, the respective positions of the Earth, the Moon, and the Sun led to a ground observer, located not far from the Equator having an eclipse whose duration of 7 min 4 s was almost the maximum possible. This was also the case during the 1954 eclipse, observed by an aircraft, without a window as we have discussed above, by the British astronomer D.E. Blackwell.

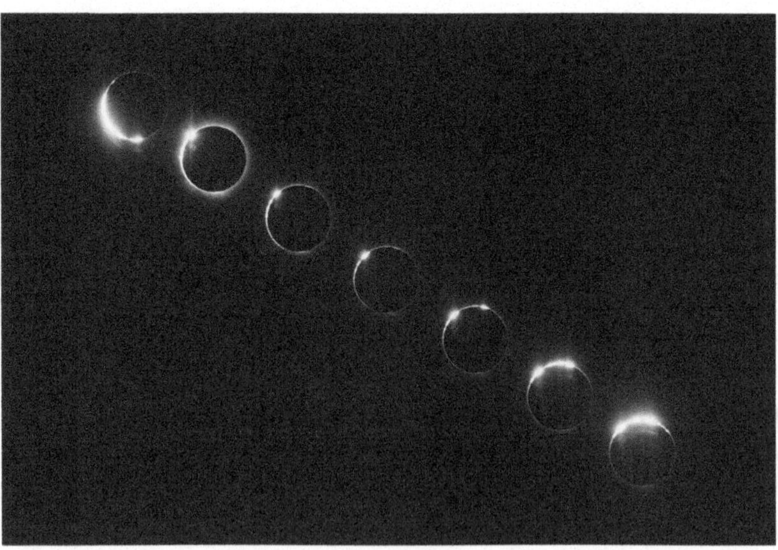

Fig. A.4 Eclipse of November 3, 2013, photographed in successive images at an altitude of 13,700 m (45,000 feet) from a Falcon aircraft over the Atlantic, 600 nautical miles SE of the Bermuda Islands. In the case of this eclipse, the distances from the Earth to the Sun and the Moon are such that at the moment of totality there remains the light of a very thin chromospheric ring. The aircraft flew perpendicular to the line of centrality, so that the duration of totality was extremely short. (Source: B. Cooper/https://launchphotography.com/)

Appendix 2: The Concorde Aircraft

The *Concorde* aircraft marked the second half of the twentieth century with an extraordinary aeronautical and technical success, coupled with an exceptionally audacious human and commercial adventure (Fig. A.5). Even if the aircraft was finally withdrawn from service in 2003 after having transported thousands of passengers, these unique successes ensured the two designing countries, France and the United Kingdom, a technical lead that allowed Europe to become, with Airbus, the leading supplier of commercial aircraft in the world.

To carry a 100 people at Mach 2 safely, crossing nearly 6000 km over the Atlantic Ocean, therefore without a close-by diversion field for a large part of the flight, required solving a number of entirely new problems. One can appreciate them by noting that in the mid-1960s, when the design of the *Concorde* was decided, only a completely secret aircraft, the Lockeed SR-71, was capable of long-duration supersonic cruise... for a crew of two people.

Let us list here some of these problems, as well as the solutions that have been found.

- The wing. It is necessary to design a wing capable of the necessary lift, allowing flight at low speed (takeoff and landing) as well as supersonic flight. The modified delta wing, called a gothic wing, provides sufficient volume for fuel tanks. Its low lift at low speed requires the aircraft to land very nose up.
- The engines. The engines must be able to provide the thrust necessary, while operating with air injected at very variable speeds, which significantly

© Springer International Publishing Switzerland 2025

P. Léna, S. Koutchmy, *Eclipsed Suns, the Solar Corona and Exoplanets*, Astronomers' Universe, https://doi.org/10.1007/978-3-031-92199-5

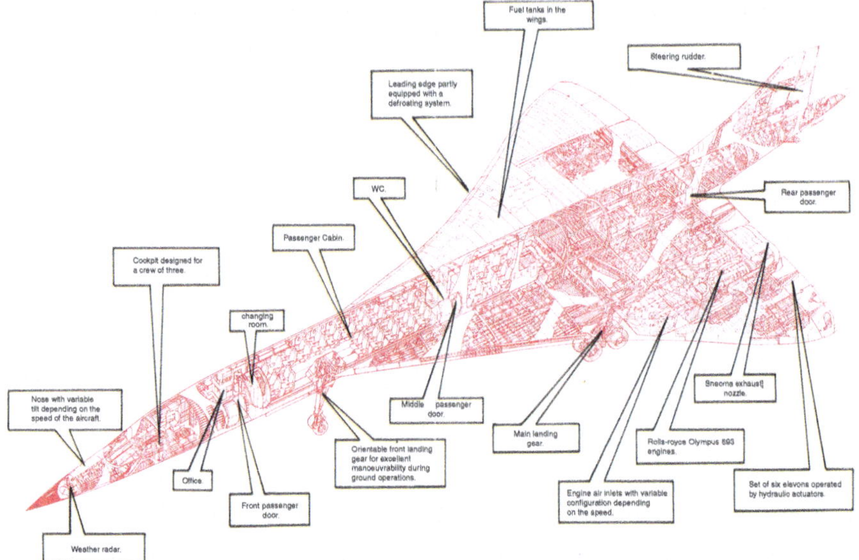

Fig. A.5 The main elements of the commercial series *Concorde* aircraft (Source: *Aviation Magazine International* & J. Pérard)

changes the combustion conditions in the reactors: the air injected at supersonic cruise at Mach 2 must become subsonic in the engine. The entry of air into the reactor is controlled by adjustable flaps, which must avoid instability (called "pumping" of a reactor) through constant control.

- The afterburner. The afterburner phase (or "reheat") increases the thrust in certain critical phases of flight: takeoff, breaking the sound barrier (from Mach 0.97 and beyond to Mach 1.7). Fuel is then injected into the engine exhaust gases, where it ignites. This use, very costly in fuel, must remain limited to the essential and cannot operate in cruise.
- The flight controls. The length of the aircraft making the use of long control cables too complex, electric controls, with motors controlled remotely by an electrical circuit, are installed close to the element to act upon, for the first time on a commercial aircraft.
- The brakes. Carbon-disc brakes, equipped with a servo, save a significant mass, prevent skidding, and reduce the stopping distance.
- The material. The aluminum alloy with which the aircraft is built must withstand a significant increase in temperature in supersonic flight: more than 100 °C on the fuselage and the leading edge of the wings. The parts are machined directly from a solid block.

All these innovations, and many others, were tested on the ground and in flight, conducted in France by André Turcat and in the United Kingdom by his counterpart Brian Trubshaw, before the production aircraft adopted them.

Many of them later found a direct application in the design of subsonic Airbus aircraft. It was their successful combination that allowed the line production of the commercial *Concorde* to meet the assigned specifications. The reader can refer to the many references that detail the *Concorde* adventure.[1]

In 2022, while the commercial airline fleet in the world is only subsonic, many researchers and industrialists continue to study supersonic commercial flight for a more or less distant future. For example, the European consortium EADS, which manufactures Airbus, presented in 2011 a project for a hypersonic aircraft, the ZEHST (*Zero Emission Hyper Sonic Transportation*). Flying at Mach 4 at an altitude of 32,000 m (105,000 feet), twice that of *Concorde*, it could carry a hundred passengers from Paris to Tokyo in an hour and a half! In 2018, the company Boom Supersonic, created in Colorado (United States), announced the project of the commercial supersonic aircraft *Overture*, with its demonstrator XB-1. The aircraft would carry up to 80 passengers over 4250 nautical miles—or 7900 km—at a Mach number of 1.7 and burning a "100% sustainable" fuel. It looks a lot like the *Concorde*.[2] Will we one day see these aircraft beat the record of a totally eclipsed Sun for 74 min?

Higher, faster, more air passengers… Is it really reasonable, "sustainable", or "supportable", as we say today? The world has changed, and humanity must learn a new way of inhabiting its planet and using its resources with more frugality and intelligence. We must dream again, but of different wonders…

---

[1] https://en.wikipedia.org/wiki/Concorde.

[2] https://boomsupersonic.com/overture.

Appendix 3: Some Data Related to the Sun and Its Atmosphere

The Sun and the Earth

	Sun	Earth	Ration Sun/Earth
Mass ($\times 10^{24}$ kg)	1,988,500	5.9724	333,000
Gravitation GM ($\times 10^6$ km^3/s^2)	132,712	0.3986	333,000
Volume ($\times 10^{12}$ km^3)	1,412,000	1.083	1,304,000
Mean radius (km)	695,700	6 371	109.2
Average specific mass (kg/m^3)	1408	5 514	0.255
Intensity of surface gravity (m/s^2)	274.0	9.78	28.0
Escape velocity (km/s)	617.6	11.19	55.2
Visual magnitude at a distance of 1 AU	-26.74	-3.86	–
Spectral type of the Sun	G2 V		

Properties of Matter at the Sun's Center

Central pressure	2.77×10^{11} bar (2.477×10^{17} g/cm^2)
Central temperature	1.571×10^7 K
Central specific mass	1.622×10^5 kg/m^3 (1.622×10^2 g/cm^3)

Parameters of Rotation and Orbits

	Sun	Earth	Ratio Sun/Earth
Sideral period of rotation (h)	60,912	239,345	25,449
Obliquity on ecliptic (°)	7.25	23.44	0.09
Relative velocity w.r.t. the stars (km/s)	19.4		

© Springer International Publishing Switzerland 2025

P. Léna, S. Koutchmy, *Eclipsed Suns, the Solar Corona and Exoplanets*, Astronomers' Universe,

https://doi.org/10.1007/978-3-031-92199-5

The Corona of the Sun

Mass of solar corona at a minimum of activity	3×10^{15} kg (about 1/1000 of the mass of the Earth's atmosphere); this mass varies by about a factor of 3 depending of the activity cycle
Time for a complete renewal of the corona	About 1 week
Average flux of solar wind at 1 AU	10^8 particles/cm²/s
Density	3 particles/cm³
Average velocity	330 km/s

Magnetic Fields in the Sun

$(1 \text{ Tesla} = 10^4 \text{ Gauss})$

Typical Values of the Magnetic Field in Different Coronal Regions and Structures

Polar regions (average absolute value)	1–2 G
Sunspots in umbra (unipolar vertical field)	3000 G
Sunspots in penumbra (horizontal field)	1500 G
Prominences	10–100 G
Facular plages	200 G
Concentrated elements	1500 G
Chromospheric network (average)	25 G
Concentrated elements	1000 G
Solar corona (extrapolations)	From 1000–0.1 G, depending on regions and radial distance

The Atmosphere of the Sun

Gaseous pressure in the photosphere (optical depth = 1 or height = 0 km)	125 mb
Effective temperature	5772 K
Minimal temperature of the photosphere, altitude 1000 km	4400 K
Temperature of the photosphere, altitude 0 km	6600 K
Average temperature of the chromosphere	About 30,000 K
Scale height in the photosphere	70–100 km
Duration of the sunspots' cycle of activity	11.4 years

Composition of the Photosphere

Abundance of the main chemical elements	Hydrogen 91%, helium 8%
Other elements (parts per million)	Oxygen 774, carbon 330, neon 112, nitrogen102, iron 43, magnesium 35, silicon 32, sulfur 15

Websites

https://www.ina.fr/contenus-editoriaux/articles-editoriaux/30-juin-1973-le-concorde-poursuit-une-eclipse-solaire/

Air and Space Museum (France) and the 1973 Eclipse: https://www.youtube.com/watch?v=oYC3Tudu5DA&list=PLVdvbycmeDGzs-5mYjcBE_g2pGeI9A2bF

http://www.astroexplorer.org/

Eclipse expeditions (Williams College, United States): https://web.williams.edu/Astronomy/eclipse/

Ephemerides, historical maps and eclipse expeditions: http://xjubier.free.fr/en/site_pages/SolarEclipsesGoogleMaps.html

Exoplanets Encyclopedia (Paris Observatory), with a constant updating: http://exoplanet.eu/

Data base of images for astronomical objects (American Astronomical Society): http://worldwidetelescope.org/webclient/

© Springer International Publishing Switzerland 2025

P. Léna, S. Koutchmy, *Eclipsed Suns, the Solar Corona and Exoplanets*, Astronomers' Universe,

https://doi.org/10.1007/978-3-031-92199-5

References

Aime, C., et al. 2013. The Fresnel diffraction: A story of light and darkness. *European Astronomical Society Publications Series* 59:37–58.

Allen, C. W. 1946. The spectrum of the Corona at the eclipse of 1940 October 1. *Monthly Notices of the Royal Astronomical Society* 106:137.

Avice M. et al. 2024. Objectivité dans la représentation des éclipses au milieu du XIXe siècle. Personal communication to the authors.

Baikousis, C., and M. O. Magnasco. 2008. Is an eclipse describedf in the Odyssee? *Proceedings of the National Academy of Sciences of the United States of America* 105:8823–8828.

Balbus, S. A. 2014. Dynamical, biological and anthropic consequences of equal lunar and solar angular diamete. *Proceedings of the Royal Society* 470:2168.

Battams, K., et al. 2020. Parker solar probe observations of a dust trail in the orbit of (3200) Phaethon. *The Astrophysical Journal* 264:64.

Benisty, M., et al. 2021. A circumplanetary disc around PDS70. *The Astrophysical Journal Letters* 916:15.

Blackwell, D. E. 1955. A study of the outer corona from a high-altitude aircraft at the eclipse of 1954 June 30. *Monthly Notices of the Royal Astronomical Society* 115:629–649.

Bonnet, R. M. 1992. *Les Horizons chimériques*. Paris: Dunod.

Bonnet, R. M. 2000. *Les premières expériences françaises de physique solaire dans l'espace*, 59. Paris: European Space Agency.

CERIMES. 1973. Eclipse 73. Directed by G. Dassonville, P. Léna and J. Rösch.

Charney. 1979. *Carbon dioxide and climate. A scientific assessment*. Washington: National Academies of Sciences.

Defrère, D., et al. 2011. Hot exozodiacal dust resolved around Vega with IOTA/IONIC. *Astronomy & Astrophysics* 534:A5.

© Springer International Publishing Switzerland 2025

P. Léna, S. Koutchmy, *Eclipsed Suns, the Solar Corona and Exoplanets*, Astronomers' Universe,

https://doi.org/10.1007/978-3-031-92199-5

Delannée, C., et al. 1998. Coronal plasmoid dynamics. I. Dissipative MHD approach. *Astronomy & Astrophysics* 329:1111–1118.

Dolci, W. W. 1997. Milestones in airborne astronomy: from the 1920s to the present. *SAE Transactions* 106:13.

Dyson, F. W., and A. S. Eddington. 1920. A determination of the deflection of light by the sun's gravitational field, from the observations made at the total eclipse of May 29, 1919. *Philosophical Transactions of the Royal Society of London* 220:291–333.

Eddy, J. A. 1972. *This dark brighness that falls from the stars*. Cham: Springer.

Eddy, J. A. 1979. Edison the scientist. *Applied Optics* 18:3737–3750.

ESA. 2021. *Selection of the science themes for the Voyage 2050 plan*. Paris: European Space Agency.

Filippov, A., and S. Koutchmy. 2002. About the prominence heating mechanisms during its eruptive phase. *Solar Physics* 208:283–295.

Grimm, S., et al. 2018. The nature of TRAPPIST-1 exoplanets. *Astronomy & Astrophysics* 613:A68.

Galicher R., Mazoyer J. Compte-Rendus. *Physique*. 2023. 24 (S2), 69–113.

Guillermier, P., and S. Koutchmy. 1999. *Total eclipses. Science, observations, myths and legends*. Cham: Springer.

Haffert, S. Y., et al. 2019. Two accreting protoplanets around the young star PDS70. *Nature Astronomy* 3:749.

Harju, J., et al. 2017. Radio interferometric observation of an asteroid occulation. *The Astronomical Journal* 156:155.

Hecht, G. 2002. *Optics*. San Francisco: Addison Wesley.

International-Consortium. 2020. Atomic data for plasma spectroscopy: The CHIANTI database, improvements and challenges. *Atoms* 8:46.

Kalas P. et al. 2013. STIS Coronagraphic Imaging of Fomalhaut: Main Belt Structure and the Orbit of Fomalhaut b. *The Astrophysical Journal* 775:56.

Kenneth, O. 1982. *Concorde: New shape in the sky*. London: Janes.

Kenneth, Owen. 2002. In concorde, an ICBH witness seminar. In *Concorde, an ICBH witness seminar*. London: Institute of Contemporary British History.

Keppler M. et al. 2018. Discovery of a planetary-mass companion within the gap of the transition disk around PDS70. *Astronomy & Astrophysics* 617:A44.

Koutchmy, S. 1972. On the demonstration and interpretation of external diffuse reinforcements in the solar corona of March 7, 1970. *Astronomy & Astrophysics* 226:103–107.

Koutchmy, S. 1976. L'étude de la couronne blanche à bord de Concorde 001 au cours de l'éclipse totale de Soleil du 30 juin 1973. *L'Astronomie* 89:149–157.

Koutchmy, S., and F. Magnant. 1973. On the observation of the F-Corona in the vicinity of the solar limb. *The Astrophysical Journal* 186:671–677.

Koutchmy, S., and M. Molodensky. 1992. 3D image of the solar corona from W-L observations of the 1991 eclipse. *Nature* 360:717.

Koutchmy, S., and G. Nikolsky. 1983. The night sky from salyut. *Sky and Telescope* 65:23–25.

Koutchmy, S., I. D. Zugzhda, and V. Locans. 1983. Short period coronal oscillations - Observation and interpretation. *Astronomy & Astrophysics* 120:185–191.

Koutchmy, S., et al. 1973. Rapid variations observed during the total eclipse of the Sun on June 30, 1973. *Nature* 246:414–415.

Koutchmy, S., et al. 1974. Photométrie photographique de la couronne solaire. *Solar Physics* 35:369–375.

Koutchmy, S., et al. 1978. Photometric analysis of the solar corona. *Astronomy & Astrophysics* 69:35–42.

Koutchmy, S., et al. 1993. CFHT eclipse observation of the very fine-scale solar corona. *Astronomy & Astrophysics* 281:249–257.

Koutchmy, S., et al. 1995. The finest white-light coronal features. *Solar Physics* 148:169–172.

Koutchmy, S., et al. 2019. New deep coronal spectra from the 2017 total solar eclipse. *Astronomy & Astrophysics* 632:A86.

Kral, Q., et al. 2017. Exozodiacal clouds: Hot and warm dust around main sequence stars. *The Astronomy and Astrophysics Review* 2:69–111.

Lamy, P., et al. 1992. No evidence of a circumsolar dust ring from the infrared observations of the 1991 solar eclipse. *Science* 257:1377.

Lamy, P., and L. Gilardy. 2022. Observation of the solar F-corona from space. *Space Science Reviews* 218:53.

Landry-Deron, I. 2013. *La Chine des Ming et de Matteo Ricci (1552-1610): le premier dialogue des savoirs avec l'Europe*. Paris: Editions du Cerf/institut Ricci.

Léna, P. 2018. The Panhard Gazogene mission in French Africa (1929). Bulletin de l'Association Les Doyennes de Panhard & Levassor.

Léna, P. 2020. *Astronomy's quest for sharp images. From blurred pictures to the very large telescope*. Cham: Springer.

Lescure, J. 1979. Comportement vocal des amphibiens et des oiseaux au Surinam. Soleil est mort: l'éclipse totale du 30 Juin 1973, Francillon G. & Menget P., 91–103.

Manel, J.-P. 1971. *La grande aventure de Concorde*. Paris: Presses Pocket.

Mann, I. 2016. Comets as a possible source of nanodust in the Solar System cloud and in planetary debris discs. *Philosophical Transactions of the Royal Society* 375:254.

Mann, I., and A. Czechowski. 2021. Dust observations from Parker Solar Probe: Dust ejection from the inner Solar System. *Astronomy & Astrophysics* 650:A29.

Masson, M. 2009. *Matteo Ricci, un jésuite en Chine*. Paris: Facultés jésuites de Paris.

McComas, D. J., et al. 2002. The three-dimenbsional solar wind around solar maximum. *Geophysical Research Letters* 30:17136.

Mercier, R., and J.M. Pasachoff. 1969. Ninety minutes of totality. Sky and Telescope.

Michels, D. J., et al. 1982. Observations of a comet on collision course with the Sun. *Science* 215:26.

Morris, L. R. 2013. Biplan Voisin: l'éclipse de 1912, la naissance de l'astronomie en avion. L'Astronomie. Société Astronomique de France.

Mulkin, B. 1981. In flight: The story of Los Alamos eclipse expeditions. *Los Alamos Science* 2:39.

NAS. 2023. *Origins, worlds and life: A decadal strategy for planetary science and astrobiology (2023-2032)*. Washington: National Academies of Sciences, Engineering and Medicine.

Nikolski, G., et al. 1985. Photographic observations of the inner zodiacal light aboard Salyut 7. In *Astrophysics and space science library*, 7–10. Dordrecht: Reidel Publishing.

Nobrega-Siverio, D., and F. Moreno-Insertis. 2022. A 2D model for coronal bright points: Association with spicules, UV bursts, surges, and EUV coronal jets. *The Astrophysical Journal Letters* 935:L21.

November, L. J., and S. Koutchmy. 1996. White-light coronal dark threads and density fine structure. *The Astrophysical Journal* 466:512.

Noyes, R. M., J. M. Beckers, and F. Low. 1968. Observational studies of the solar intensity profile in the far infrared and millimeter regions. *Solar Physics* 3:36–46.

Noyes, R. M., J. M. Beckers, and F. Low. 2023. *Origins, worlds and life: A decadal strategy for planetary science and astrobiology (2023-2032)*. Washington: National Academies of Sciences, Engineering and Medicine.

Rouan, D. 2012. Osiris et astroplane: l'astronomie infrarouge aéroportée en France. *L'Astronomie* 6:62.

Rudniskij, G. M. 2006. I.S. Shklovsky and modern radioastronomy. *Astronomy & Astrophysics Transactions* 25:363–368.

Saint-Exupery, A. 1974. *Night flight (transl. from French)*. New York: Harcourt Brace Jovanovich.

Scazay, J. R., et al. 2021. Collisional evolution of the inner zodiacal cloud. *The Planetary Science Journal* 2:185.

Shestakova, L. I. 2003. The Beta-Pictoris phenomenon near the Sun. *Astronomy & Astrophysics Transl.* 22(2):191–211.

Smart, R. L., et al. 2021. Gaia early data release 3: The Gaia catalogue of nearby stars. *Astronomy & Astrophysics* 649:A6.

Smith B.A., Terrile R. 1984. A circumstellar disk around Beta Pictoris. *Science* 226:1421–1424.

Snyder, C. W., M. Neugebauer, and U. Rao. 1963. The solar wind velocity and its correlation with cosmic-ray variations and with solar and geomagnetic activity. In *Proccedings of the 8th International Cosmic Rays Conference*, 210. Cham: Springer.

Souclet, E. 1732. *Observations mathématiques, astronomiques, géographiques, chronologiques et physiques*. Paris: Rollin.

Stellmacher, G., and S. Koutchmy. 1974. Study of low dispersion coronal spectra. *Astronomy & Astrophysics* 35:43–48.

Su, K. Y. L., et al. 2005. The Vega debris disk: A surprise from Spitzer. *The Astrophysical Journal* 628:487.

Tavabi, E., S. Koutchmy, and A. Ajabshirizadeh. 2011. Contribution to the modeling of solar spicules. *Advances in Space Research* 47 (11): 2019–2029.

Tavabi, E., et al. 2013. Increasing the fine structure visibility of the hinode SOT Ca II H filtergrams. *Solar Physics* 283:187–194.

Tornay, S. 1979. Médecine du corps, médecine du cosmos: l'éclipse chez les Nyangatom. In Soleil est mort: l'éclipse totale du 30 Juin 1973, by G. Francillon and P. Menget, 201–243.

Turcat, A. 2013. L'ombre pour toute proie. Bulletin de l'Académie des sciences, agriculture, arts et belles lettres d'Aix-en-Provence.

Turcat, A. 2020. *Concorde:essais et batailles*. Paris: Le Cherche-Midi.

Turcat, A. 2021. Dernier survol.Une soif d'apprendre. J.P.Otelli.

Vandenbussche, B., et al. 2010. The β Pictoris disk imaged by Herschel PACS and SPIRE. *Astronomy & Astrophysics* 518:L133.

Wraight, P., and M. Gasden. 1975. Dayglow of the infrared atmospheric band system of O2 during a total eclipse of the Sun. *The Journal of Atmospheric and Solar-Terrestrial Physics* 37:717–730.

Yokohama, T., and K. Shibataka. 1998. A two dimension magnetohydrodynamic simulation of chromospheric evaporation in a solar flare based on a magnetic reconnection model. *Astrophysical Journal* 494:113-116.

Zhukov, A., et al. 2000. Coronal plasmoid dynamics II. The non-stationary fine structure. *Astronomy & Astrophysics* 353:785–796.

Zirker, J., S. Koutchmy, et al. 1992. Structural changes in the solar corona during the July 1991 eclipse. *Astronomy & Astrophysics* 385:L1–L4.

Further Reading

Aschwanden, M. 2006. *Physics of the solar corona*. Berlin: Springer Science & Business Media.

Cox, A. N., ed. 2001. *Allen's astrophysical quantities*. Cham: Springer.

Golub, L., and J. M. Pasachoff. 2009. *The solar corona*. Cambridge: Cambridge University Press.

Judge, P. G., and J. A. Ionson. 2023. *The problem of coronal heating. A Rosetta stone for electrodynamic coupling in cosmic plasmas*. Cham: Springer.

Maksimovic, M. 2024. *Le grand atlas du Soleil*. Paris: Glénat.

Meyer-Vernet, N. 2007. *Basics of the solar wind*. Cambridge: Cambridge University Press.

The manufacturer's authorised representative in the EU is Springer
Nature Customer Service Centre GmbH, Europaplatz 3, 69115 Heidelberg,
Germany. If you have any concerns regarding our products, please
contact ProductSafety@springernature.com

Printed and bound by CPI Group (UK) Ltd, Croydon, CR0 4YY

03/07/2026

02155952-0001